The nature and art of workmanship

The nature and art of workmanship

David Pye

Formerly Professor of Furniture Design
Royal College of Art, London

Cambridge University Press

Cambridge
London New York New Rochelle
Melbourne Sydney

Published by the Press Syndicate of the University of Cambridge
The Pitt Building, Trumpington Street, Cambridge CB2 1RP
32 East 57th Street, New York, NY 10022, USA
296 Beaconsfield Parade, Middle Park, Melbourne 3206, Australia

Library of Congress catalogue card number: 68-12062

ISBN 0 521 06016 8 hard covers
ISBN 0 521 29356 1 paperback

First published 1968
First C.U.P. paperback edition 1978
Reprinted 1979

First printed in Great Britain at the
University Press, Cambridge
Reprinted and bound in Great Britain by
Fakenham Press Limited, Fakenham, Norfolk

To the memory of E.B. Pye

Contents

Text-figures

Plates

Acknowledgements

I am indebted to Mr Christopher Cornford for his encouragement and criticism, and to Mr Ernest Joyce and Lieut.-Col. Kellow Pye for their generous assistance in developing and printing some of the photographs.

I am indebted also to those who provided, or allowed me to take, certain of the photographs, and acknowledgement is made to them in the commentary on the plates.

Preface to the paperback edition

The Crafts

In the last ten years there has taken place quite suddenly a great increase in the practice and appreciation of the Crafts. When this book was written, twelve years ago and more, this could not be foreseen or even hoped for. Consequently some passages about the future of the Crafts in Chapter 11 can now be seen to have been far too pessimistic. 'If the Crafts survive...' I wrote, and more in the same strain. *Survive* indeed! How oddly that reads now! But other things that were written there about the Crafts still seem pertinent: about design: about the high importance of the amateurs: about the best work being done for love not for money: about the necessity of aiming high: about 'gritty pots and hairy cloth' and the travesties of rough workmanship. For the sake of these I hope my failures in the matter of prophecy may be excused. At least they are a convincing demonstration that things can change for the better, and change fast—for now there is indeed quite a strong demand for the best quality and not a discreditable one either. Some few people do make a living by supplying it, and the Crafts, by and large, are becoming decidedly viable. All that is to the contrary of what I could expect and what I wrote those few years ago.

It has not been practicable to rewrite the text for this paperback edition, beyond making minor corrections, and so Chapter 11 has been allowed to stand as it was originally written.

January 1978 D.P.

1

Design proposes. Workmanship disposes

In the years since 1945 there has been an enormous intensification of interest in Design. The word is everywhere. But there has been no corresponding interest in workmanship. Indeed there has been a decrease of interest in it. Just as the achievements of modern invention have popularly been attributed to scientists instead of to the engineers who have so often been responsible for them, so the qualities and attractions which our environment gets from its workmanship are almost invariably attributed to design.

This has not happened because the distinction between workmanship and design is a mere matter of terminology or pedantry. The distinction both in the mind of the designer and of the workman is clear. Design is what, for practical purposes, can be conveyed in words and by drawing: workmanship is what, for practical purposes, can not. In practice the designer hopes the workmanship will be good, but the workman decides whether it shall be good or not. On the workman's decision depends a great part of the quality of our environment.

Gross defects of workmanship the designer can, of course, point out and have corrected, much as a conductor can at least insist on his orchestra playing the right notes in the right order. But no conductor can make a bad orchestra play well; or, rather, it would take him years to do it; and no designer can make bad workmen produce good workmanship. The analogy between workmanship and musical performance is in fact rather close. The quality of the concert does *not* depend wholly on the score, and the quality of our environment does *not* depend on its design. The score and the design are merely the first of the essentials, and they can be nullified by the performers or the workmen.

Our environment in its visible aspect owes far more to workmanship than we realize. There is in the man-made world a whole domain of quality which is not the result of design and owes little to the designer. On the contrary, indeed, the designer is deep in its debt, for every card in his hand was put there originally by the workman. No architect could specify ashlar until a mason had perfected it and shown him that it could be done. Designers have only been able to exist by exploiting what workmen have evolved or invented.

This domain of quality is usually talked of and thought of in terms of material. We talk as though the material of itself conferred the quality. Only to name precious materials like marble, silver, ivory, ebony, is to evoke a picture of thrones and treasures. It does not evoke a picture of grey boulders on a dusty hill or logs of ebony as they really are—wet dirty lumps all shakes and splinters! Material in the raw is nothing much. Only worked material has quality, and pieces of worked material are made to show their quality by men, or put together so that together they show a quality which singly they had not. 'Good material' is a myth. English walnut is not good material. Most of the tree is leaf-mould and firewood. It is only because of workmanlike felling and converting and drying and selection and machining and setting out and cutting and fitting and assembly and finishing—particularly finishing—that a very small proportion of the tree comes to be thought of as good material; not because a designer has specified English walnut. Many people seeing a hundred pounds worth of it in a London timber yard would mistake it for rubbish, and in fact a good half of it would be: would have to be.

So it is with all other materials. In speaking of good material we are paying an unconscious tribute to the enormous strength of the traditions of workmanship still shaping the world even now (and still largely unwritten). We talk as though good material were found instead of being made. It is good only because workmanship has made it so. Good workmanship will make something better out of pinchbeck than bad will out of gold. *Corruptio optimi pessima*! Some materials promise far more than others but only the workman can bring out what they promise.

In this domain of quality our environment is deteriorating. What threatens it most is not bad workmanship. Much workmanship outside of mass-production is appallingly bad and getting worse, to be sure, and things are seen in new buildings which make one's hair rise. But at least it is easy to see what the remedies are, there, if difficult to apply them. Moreover, it is not the main danger, because it is outside the field of mass-production, and the greater part of all manufacture now is mass-production; in which, although there is some bad workmanship, much is excellent. Much of it has never been surpassed and some never equalled. The deterioration comes not because of bad workmanship in mass-production but because the range of qualities which mass-production is capable of just now is so dismally restricted; because each is so uniform and because nearly all lack depth, subtlety, overtones, variegation, diversity, or whatever you choose to call that which distinguishes the workmanship of a Stradivarius violin, or something much rougher like a modern ring-net boat. The workmanship of a motor-car is something to marvel at, but a street full of parked cars is jejune and depressing; as if the same short tune of clear unmodulated notes were being endlessly repeated. A harbour full of fishing-boats is another matter.

Why do we accept this as inevitable? We made it so and we can unmake it. Unless workmanship comes to be understood and appreciated for the art it is, our environment will lose much of the quality it still retains.

2

The workmanship of risk and the workmanship of certainty

Workmanship of the better sort is called, in an honorific way, craftsmanship. Nobody, however, is prepared to say where craftsmanship ends and ordinary manufacture begins. It is impossible to find a generally satisfactory definition for it in face of all the strange shibboleths and prejudices about it which are acrimoniously maintained. It is a word to start an argument with.

There are people who say they would like to see the last of craftsmanship because, as they conceive of it, it is essentially backward-looking and opposed to the new technology which the world must now depend on. For these people craftsmanship is at best an affair of hobbies in garden sheds; just as for them art is an affair of things in galleries. There are many people who see craftsmanship as the source of a valuable ingredient of civilization. There are also people who tend to believe that craftsmanship has a deep spiritual value of a somewhat mystical kind.

If I must ascribe a meaning to the word craftsmanship, I shall say as a first approximation that it means simply workmanship using any kind of technique or apparatus, in which the quality of the result is not predetermined, but depends on the judgement, dexterity and care which the maker exercises as he works. The essential idea is that the quality of the result is continually at risk during the process of making; and so I shall call this kind of workmanship 'The workmanship of risk': an uncouth phrase, but at least descriptive.

It may be mentioned in passing that in workmanship the care counts for more than the judgement and dexterity; though care may well become habitual and unconscious.

With the workmanship of risk we may contrast the workmanship of certainty, always to be found in quantity production, and found in its pure state in full automation. In workmanship of this sort the quality of the result is exactly predetermined before a single saleable thing is made. In less developed forms of it the result of each operation done during production is predetermined.

The workmanship of certainty has been in occasional use in undeveloped and embryonic forms since the Middle Ages and I should suppose from much earlier times, but all the works of men which have been most admired since the beginning of history have been made by the workmanship of risk, the last three or four generations only excepted. The techniques to which the workmanship of certainty can be economically applied are not nearly so diverse as those used by the workmanship of risk. It is certain that when the workmanship of certainty remakes our whole environment, as it is bound now to do, it will also change the visible quality of it. In some of the following chapters I shall discuss what may be lost and gained.

The most typical and familiar example of the workmanship of risk is writing with a pen, and of the workmanship of certainty, modern printing. The first thing to be observed about printing, or any other representative example of the workmanship of certainty, is that it originally involves more of judgement, dexterity, and care than writing does, not less: for the type had to be carved out of metal by hand in the first instance before any could be cast; and the compositor of all people has to work carefully: and so on. But all this judgement, dexterity and care has been concentrated and stored up before the actual printing starts. Once it does start, the stored-up capital is drawn on and the newspapers come pouring out in an absolutely predetermined form with no possibility of variation between them, by virtue of the exacting work put in beforehand in making and preparing the plant which does the work: and making not only the plant but the tools, patterns, prototypes and jigs which enabled the plant to be built, and all of which had to be made by the workmanship of risk.

Typewriting represents an intermediate form of workmanship, that of limited risk. You can spoil the page in innumerable ways, but the N's will never look like U's, and, however ugly the typing, it will almost necessarily be legible. All workmen using the workmanship of risk are constantly devising ways to limit the risk by using such things as jigs and templates. If you want to draw a straight line with your pen, you do not go at it freehand, but use a ruler, that is to say, a jig. There is still a risk of blots and kinks, but less risk. You could even do your writing with a stencil, a more exacting jig, but it would be slow.

Speed in production is usually the purpose of the workmanship of certainty but it is not always. Machine tools, which, once set up, perform one operation, such for instance as cutting a slot, in an absolutely predetermined form, are often used simply for the sake of accuracy, and not at all to save time or labour. Thus in the course of doing a job by the workmanship of risk a workman will be working freehand with a hand tool at one moment and will resort to a machine tool a few minutes later.

In fact the workmanship of risk in most trades is hardly ever seen, and has hardly ever been known, in a pure form, considering the ancient use of templates, jigs, machines and other shape-determining systems,* which reduce risk. Yet in principle the distinction between the two different kinds of workmanship is clear and turns on the question: 'is the result predetermined and unalterable once production begins?'

Bolts can be made by an automatic machine which when fed with blanks repeatedly performs a set sequence of operations and turns out hundreds of finished bolts without anyone even having to look at it. In full automation much the same can be said of more complex products, substituting the words 'automated factory' for 'automatic machine'. But the workmanship of certainty is still often applied in a less developed form where the product is made by a planned sequence of operations, each of which has to be started and stopped by the operative, but with the result of each one predetermined and outside his control. There are also hybrid forms of production where some of the operations have predetermined results and some are performed by the workmanship of risk. The craft-based industries, so called, work like this.

Yet it is not difficult to decide which category any given piece of work falls into. An operative, applying the workmanship of certainty, cannot spoil the job. A workman using the workmanship of risk assisted by no matter what machine-tools and jigs, can do so at almost any minute. That is the essential difference. The risk is real.

But there is much more in workmanship than not spoiling the job, just as there is more in music than playing the right notes.

There is something about the workmanship of risk, or its results; or

* Shape-determining systems are discussed in my book *The Nature and Aesthetics of Design* (Barrie and Jenkins, 1978), especially in the chapters on Techniques and on 'Useless work'.

something associated with it; which has been long and widely valued. What is it, and how can it be continued? That is one of the principal questions which I hope this book may answer: and answer factually rather than with a series of emotive noises such as protagonists of craftsmanship have too often made instead of answering it.

It is obvious that the workmanship of risk is not always or necessarily valuable. In many contexts it is an utter waste of time. It can produce things of the worst imaginable quality. It is often expensive. From time to time it had doubtless been practised effectively by people of the utmost depravity.

It is equally obvious that not all of it is in jeopardy: for the whole range of modern technics is based on it. Nothing can be made in quantity unless tools, jigs, and prototypes, both of the product and the plant to produce it, have been made first and made singly.

It is fairly certain that the workmanship of risk will seldom or never again be used for producing things in quantity as distinct from making the apparatus for doing so; the apparatus which predetermines the quality of the product. But it is just as certain that a few things will continue to be specially made simply because people will continue to demand individuality in their possessions and will not be content with standardization everywhere. The danger is not that the workmanship of risk will die out altogether but rather that, from want of theory, and thence lack of standards, its possibilities will be neglected and inferior forms of it will be taken for granted and accepted.

There was once a time when the workmanship of certainty, in the form colloquially called 'mass-production', generally made things of worse quality than the best that could be done by the workmanship of risk—colloquially called 'hand-made'. That is far from true now. The workmanship of a standard bolt or nut, or a glass or polythene bottle, a tobacco-tin or an electric-light bulb, is as good as it could possibly be. The workmanship of risk has no exclusive prerogative of quality. What it has exclusively is an immensely various range of qualities, without which at its command the art of design becomes arid and impoverished.

A fair measure of the aesthetic richness, delicacy and subtlety of the workmanship of risk, as against that of certainty, is given by comparing the contents of, say, the British Museum with those of a good department store.

Nearly everything in the Museum has been made by the workmanship of risk, most things in the store by the workmanship of certainty. Yet if the two were compared in respect of the ingenuity and variety of the devices represented in them the Museum would seem infantile. At the present moment we are more fond of the ingenuity than the qualities. But without losing the ingenuity we could, in places, still have the qualities if we really wanted them.

3

Is anything done by hand?

Things are usually made by a succession of different operations, and there are often alternative ways of carrying any one of them out. We can saw, for instance, with a hand-saw, an electrically driven band-saw, a frame-saw, and in other ways.

To distinguish between the different ways of carrying out an operation by classifying them as hand- or machine-work is, as we shall see, all but meaningless. But if we make an estimate of the degree of risk to the quality of the result which is involved in each we have a real and useful basis for comparison between them. Let us take two extreme examples: (*A*) A dentist drilling a tooth with an electrically driven drill. (*B*) A man drilling a piece of wood with a hand-driven wheelbrace, using a twist-drill and a jig. *A* is a machine-operation and *B* is a hand-operation: or, if you like, we will say that both are machine-operations. Operation *A* which the dentist does with a power-driven machine-tool involves 100 per cent risk (and there is no man that it lies in his mouth to deny it!) but operation *B* merely involves a five per cent risk or so, and only that because, if the hand-workman is fool enough, he may break the drill. Otherwise he has only to keep winding the handle and the result is a certainty. The source of power is completely irrelevant to the risk. The power tool may need far more care, judgement and dexterity in its use than the hand-driven one.

Let us consider some possible definitions of handicraft, or hand-work, or work done by hand. 'Done by hand' as distinct from work done by what? By tools? Some things actually can be made without tools it is true, but the definition is going to be rather exclusive for it will take in baskets and coiled pottery, and that is about all! Let us try something wider and say 'done by hand-tools as distinct from work done by machines'. Now we shall have to define 'machine' so as to exclude a hand-loom, a brace and bit, a wheel-brace, a potter's wheel and the other machines and tools which belong to what is generally accepted as hand-work. So that will not do either, unless we propose to flout the ordinary usage of mechanics: which on the subject of machinery seems a trifle risky.

Suppose that we try 'As distinct from power-driven machine tools'. Now we are faced with having to agree that the distinction between handicraft and not-handicraft has nothing to do with the result of handicraft—the thing made: for no one can possibly tell by looking at something turned, whether it was made on a power-driven, foot-driven, boy- or donkey-driven lathe. And then again, if we hold to this definition, do we say 'made *entirely* without the use of power-driven machine tools' or do we say 'made *partly* without...'? If we say 'entirely', then all the carpentry, joinery, and cabinet-making of the last hundred years is excluded, pretty nearly: indeed for longer than that. Louis Mumford remarks* (in a different context) that '...If power machinery be a criterion, the modern industrial revolution began in the twelfth century and was in full swing by the fifteenth.' The sawmill is a very ancient thing and so, of course, is the water-driven hammer.

But if we take the other course and say 'Partly without power-driven machine-tools' we include in handicraft most of the worst products of cheap quantity-production. Perhaps we can save the situation yet, by putting in a disclaimer and saying 'made *singly*, partly without power-driven machine-tools'. But now how do we know he hasn't made two of them and kept quiet about it? There is nothing about the product, the thing made, to tell us. And if we say 'in small numbers' why, exactly, do we include six and exclude seven or such-like? It sounds more like an expedient than a definition.

Suppose that we make a last attempt, shape a different course altogether, and say 'made by hand-guided tools, whether power-driven machine-tools or not'. By so doing we have written off every kind of drill, lathe, plane, and shooting board, all of which are shape-determining systems. So we shall now have to qualify the definition to include these tools which are only in part hand-guided; and then we shall have to try to exclude whatever machines we do not happen to fancy, from the same group.

Or shall we? Is it not time to give up and admit that we are trying to define in the language of technology a term which is not technical?

'Handicraft 'and 'Hand-made' are historical or social terms, not technical ones. Their ordinary usage nowadays seems to refer to workmanship *of any kind* which could have been found before the Industrial Revolution.

* *Technics and Civilization.*

Mumford, extending a conception of Patrick Geddes's, described* three phases in the development of European economy and technics, each phase having a distinct pattern of economy and culture and a 'technical complex' of its own, which might be roughly indicated by referring to its principal materials and sources of power. The Eotechnic phase was reckoned to extend from about A.D. 1000 to 1760, and was a 'water-and-wood complex'. The Paleotechnic phase, of the Industrial Revolution, was a 'coal-and-iron complex', and the Neotechnic phase of our own day, which succeeded it, is an 'Electricity-and-alloy complex' (Mumford was writing in 1930).

The essential ideas in his conception are, I think, first: that the Eotechnic phase contained, not so much the seeds, as the nine-month embryo of the Industrial Revolution; for all the pre-requisite ideas, devices and techniques for it were already in being before it came about. Secondly: that the different phases 'interpenetrated and overlapped'. That is to say that, just as the technical features of the Paleotechnic phase, such as quantity-production and the workmanship of certainty, were in being quite early in the Eotechnic phase, so did techniques and devices characteristic of that phase persist through the Paleotechnic phase and even into our own day. I lately saw a wooden barrel (Eotechnic) with, beside it, a galvanized steel bucket (Paleotechnic) and a thermoplastic watering-can (Neotechnic). As for the workmanship of certainty having appeared during the Eotechnic phase, to quote but two examples: the monk Theophilus in the eleventh century gave a detailed description of punches and stamps for producing quantities of standardized ornaments in gold and silver,† and in or about 1294 a smith called Thomas, from Leighton Buzzard, used stamping dies for forging standardized ornamental features for the grille of Eleanor of Castile's tomb in Westminster Abbey, which still exists.‡ It may be that in its earliest manifestations the workmanship of certainty was used for the quantity-production of ornaments more often than for utilitarian purposes.

Now the current idea of handicraft and the hand-made has been deeply coloured by the Arts and Crafts movement; and that became a movement of protest against the workmanship and aesthetics of the Industrial Revolution, which it contrasted with handicraft. As a result, I think, the idea has become accepted that before the Industrial Revolution everything was made

* *Technics and Civilization.*

† See H. Wilson, *Silverwork and Jewellery* (1903).

‡ See H. R. Schubert, *History of the British Iron and Steel Industry.*

without machines. This was certainly not William Morris's idea. He did not consider that handicraft flourished after the Middle Ages. But the fairly common error of supposing a complete break and opposition between the 'machine-made' workmanship of the Industrial Revolution and the 'hand-made' workmanship of the Eotechnic phase immediately preceding it is presumably traceable partly to a misunderstanding of Morris.

It seems fairly clear that to Morris himself handicraft meant primarily work without division of labour, which made the workman 'a mere part of a machine'. During the medieval period, he says, 'there was little or no division of labour, and what machinery was used was simply of the nature of a multiplied tool, a help to the workman's hand-labour and not a supplanter of it. The workman worked for himself and not for any capitalistic employer and he was accordingly master of his work and his time. This was the period of pure handicraft.'* It will be noted that for him handicraft did not exclude the use of machines and that the word had strong social and historical implications. It was not a word referring to any definable technique.

In this book there is no need for us to go into the question of whether Morris's beliefs about the Middle Ages are true.

One contributory cause of present confusions of thought about hand-work and craftsmanship is perhaps that people have generalized about it who did not know, or did not think enough about, the way tools do actually work. I am inclined to propose that the term hand-work should be confined to the work of a hand and an unguided tool; but that is an extremely restrictive definition. I do not think any wood-working tool can be properly said to be unguided after the moment when it enters the wood. They all cut their own jig as they work and sometimes a pretty exact one, as with a paring-chisel or a scribing-gouge. Workmanship in different trades differs so widely in its basis as well as its practice, that the only common factor and the only means of generalization in all the different branches of craftsmanship is the element of risk we have discussed.

The extreme cases of the workmanship of risk are those where a tool is held in the hand and no jig or any other determining system is there to guide it. Very few things can properly be said to have been made by hand, but, if there are any operations involving a tool which may legitimately be called hand-work, then perhaps these are they. Writing and sewing are examples.

* 'The Revival of Handicraft', *Fortnightly Review*, November 1888.

4

Quality in workmanship

We have given some account of workmanship. Now let us consider some of the epithets which are commonly applied to it. There are, I think, four which we should examine: 'good' and 'bad', 'precise' and 'rough'. It is usual to equate 'good' with 'precise' and 'bad' with 'rough'. To do so is false. Rough workmanship may be excellent while precise may be bad.

The goodness or badness of workmanship is judged by two different criteria: soundness and comeliness. Soundness implies the ability to transmit and resist forces as the designer intended; there must be no hidden flaws or weak places. Comeliness implies the ability to give that aesthetic expression which the designer intended, or to add to it. Thus the quality of workmanship is judged in either case by reference to the designer's intention, just as the quality of an instrumentalist's playing is judged by reference to the composer's.

In some cases precision is necessary to soundness, but in many others it is not, and rough workmanship will do the job just as well. In some cases precision is necessary to the intended aesthetic expression but in others it is not and, on the contrary, rough workmanship is essential to it.

All workmanship, as we shall see, is approximation, to a greater or less degree. A designer may perfectly well expect and intend the roughest of approximation. Just as a composer by a notation like 'Con brio' may indicate how he wants to be played, so may a designer. If, on the other hand, the designer intends precision and gets it in the main, but finds it interrupted by passages of approximation he never intended, then the effect will be horrible: and this is bad workmanship. Good workmanship is that which carries out or improves upon the intended design. Bad workmanship is what fails to do so and thwarts the design.

All workmanship is approximation. There are in the world of manufacture, and not only in that of metaphysics, certain Ideas of which the things we make are necessarily imperfect copies. Nothing has ever been square because nothing has ever been straight, nor has anything been flat, nor spherical, cylindrical, cubical.

Socrates, in the *Phaedo*, maintains that the idea of absolute equality is suggested to us by the sight of things which appear to be approximately equal, because they remind us of something our souls knew before we were born. A similar contention could of course be made about absolute flatness or straightness. I prefer another explanation for I do not think there can be much doubt how we have arrived at the idea of an absolutely flat surface when nothing flat exists. Whenever we make something 'flat' and find it is not flat enough, we always find that by taking more trouble we can make it still flatter: or we have always been able to do so hitherto: and so we find it easy to imagine we are approximating to a perfect flatness which it is just beyond our powers or patience to reach.

Unless we accept Plato's explanation and postulate a primordial inborn memory of ideal forms, our whole notion of geometrical perfection must have been built up by this sort of extrapolation.

Beyond this approximation to an unattainable geometrical ideal there is a second order of approximation to mere regularity. We do not always insist on exact duplication, or symmetry, or evenness of section, or fairness of a curve, or repetition of a unit.

This kind of approximation may be done deliberately, as it is for instance in the asymmetrical weaving of an essentially symmetrical pattern in some oriental rugs, for magical reasons; or it may be done as making a virtue of necessity where the desire or need for economy prompts us to rough workmanship. But, whatever reason we may give for it, in all such cases the workman admits to the work an element of the unaccountable and unstudied: of improvisation: either deliberately or because he has not the time or ability to prevent it.

Now a design is in effect a statement of the ideal form of the thing to be made, to which the workman will approximate in a greater or less degree. In a designer's drawing all joints fit perfectly! If the designer wants precise workmanship he is saying, as he shows the drawings and specification to the workman, 'This tells you how, ideally, it ought to be. Now show us how near you can get.' Or, on the other hand, he may be saying 'This is how it is supposed to be, but don't take it too literally. You know that a fairly rough job is usual in this sort of work, and that is what I should like to see.'

The trouble is that designs in so many trades are conceived in terms of combinations of simple geometrical forms. In architecture for instance it has hardly ever been otherwise. Now it happens that, as the Gestalt psychologists have demonstrated, we have a very effective inborn ability and indeed compulsion to see the straightness in all the things which are fairly straight and the triangularity in all the things which are more or less triangular.* Consequently when we see a rough-hewn baulk of timber we assume at once, without having to learn the fact, that it was 'meant to be' a rectangular prism, which it manifestly is not. Conversely, when we see what, so far as the eye can tell, *is* a perfect rectangular prism, but there happens to be a great open joint in it, we know at once that the joint was not 'meant to be' there.

Let us provisionally give the name 'perfect' workmanship to that in which the achievement seems to correspond exactly with the idea: the spherical ball-bearing appears to be exactly spherical: let us on the other hand give the name 'rough' to workmanship in which there is an evident disparity between idea and achievement. In rough work we see timbers 'roughly squared', components 'roughly lined up' and so on. In such cases we infer the idea from the achievement, the rectangular prism from the roughly squared timber.

The workman's achievement may differ from the idea for three quite separate reasons: it may do so because he intends that it shall, it may do so because he has not time to perfect the work, and finally it may do so because he has not enough knowledge, patience or dexterity to perfect it. The last of these reasons is the one with which every layman is familiar, and hence to the layman rough workmanship often suggests ineptitude. It is taken for granted that the man who did it must have been incapable of doing perfect work. To any workman or artist that idea seems laughable. Many of Rembrandt's drawings are rough, but not, one may safely say, because of ineptitude. But even where this is understood, the rough work, because less laborious, is, in the West, usually considered in some way inferior to the perfect. In the Far East this has not been so. In Japan the cult of a certain kind of rough workmanship has had a great following and become highly sophisticated.

In the workmanship of risk rough work is the necessary basis of perfect

* See R. Arnheim, *Art and Visual Perception*, chapter 2.

work, just as the sketch is of the picture. The first sketchy marks on the canvas may become the foundation of the picture and be buried, or they may be left standing. Similarly the first approximations of the workman may afterwards disappear as the work proceeds, or they may be left standing. For the painter and the workman it is sometimes difficult to know when to stop on the road towards perfect work, and sooner may be better than later. In the workmanship of certainty, on the other hand, there is no rough work. The perfect result is achieved directly without preliminary approximation.

In the case of open joints which we know are 'not meant to be there' we are confronted by a kind of bad workmanship which is very common. The

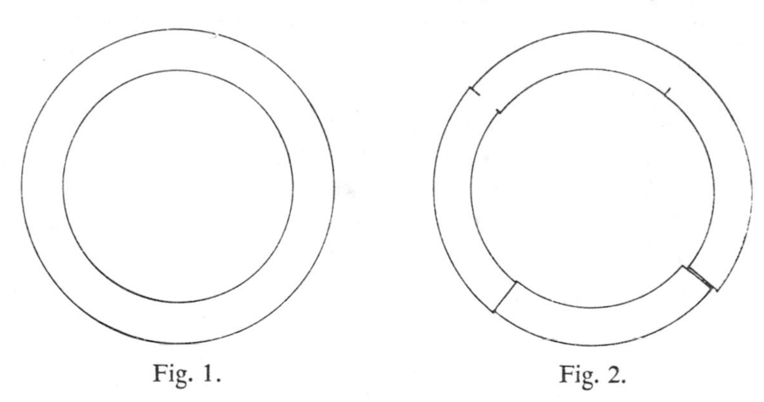

Fig. 1. Fig. 2.

workman is essentially an interpreter, and any workman's prime and overruling intention is necessarily to give a good interpretation of the design. If he fails in this he will either distort or disrupt the design, or both. If he is using a constructing technique the result of failure is usually disruptive. Let us take, by way of a test-tube example, the circular joined wooden glass-frame or picture-frame in fig. 1. The frame is a ring: a continuous form. Continuity is evidently the essence of the designer's intention. If, then, by bad workmanship the ring is broken, as in fig. 2, the continuity is interrupted and the intention flouted. Anyone who is in the smallest degree sensitive to the aesthetic intentions of design must be aware of this.

It is futile for a designer to aim at expressing continuity by means of a joined frame unless he is confident of getting expert workmanship. Otherwise he must seek the same effect by different means, either bending a single length of some suitable material into a ring, or cutting the ring out of a solid piece or else casting it. If a designer forces his intentions on workmen who, he knows, are not good enough at their job to carry them out, then he is quite as much to blame for the result as they are.

In all manufacture it is the rule rather than the exception to find that the degree of approximation in the workmanship of one piece of construction varies considerably in different parts of it and fairly often one finds rough work cheek-by-jowl with perfect work. The aesthetic success or failure of such a combination will depend on whether their being combined adds something to the design or detracts from it. If each acts as a foil to the other and sets it off, all is well, and we accept the combination without question as intentional; but, if the rough work looks like an intrusion, mere evidence of carelessness, then the job is spoilt. Very often the apportioning of perfect and rough work is decided by the workman. If, as he interprets the design, he imports a combination which the designer did not ask for, he will have to do it with discrimination and understanding.

Before we can go much farther it will be necessary to improve on the terms 'perfect' and 'rough' which were provisionally adopted; for, as we know, 'perfect' workmanship only appears to be so and is approximate, while much workmanship in which some approximation can be detected by the naked eye is certainly not what one would ordinarily call rough.

Let us then say that, where the naked eye can detect no disparity between achievement and idea, the workmanship is 'regulated' or, in cases of extreme precision, 'highly regulated'. Where slight disparities can be detected let us say that it is 'moderately free'. Where there are evident (and usually intentional) disparities, as often seen in woodcarving and calligraphy, where precise repetition is on the whole avoided, let us say the work is 'free'. And, where we should ordinarily call the work rough, let us call it rough; remembering always that rough does not necessarily imply bad.

The term 'regulated' is apt, whether applied to the workmanship of risk or to that of certainty. On the other hand, the workmanship of certainty is all but incapable of free or rough work at present; but it must be remembered that, where construction is involved in the making of something, then although the components may be made by the workmanship of certainty they will still nearly always have been assembled by the workmanship of risk. Regulated work is then possible, but, in quantity-production, bad is more probable, as in the case of the glass frame just cited.

Regulation is achieved in the workmanship of risk in three different ways, separate or combined. The first is dexterity: which means sheer adroitness in handling. The old-style shipwright with his adze can get a nearly true flat surface or a fair curve without any apparent guide, simply by coordination of hand and eye. Secondly, gradualness: the shipwright with his adze does not finish off the surface by removing handfuls of wood at each stroke, but in short light strokes taking off the wood in thin shavings. Lastly, shape-determining systems: such as jigs, forms, moulds, gauges. The variety of

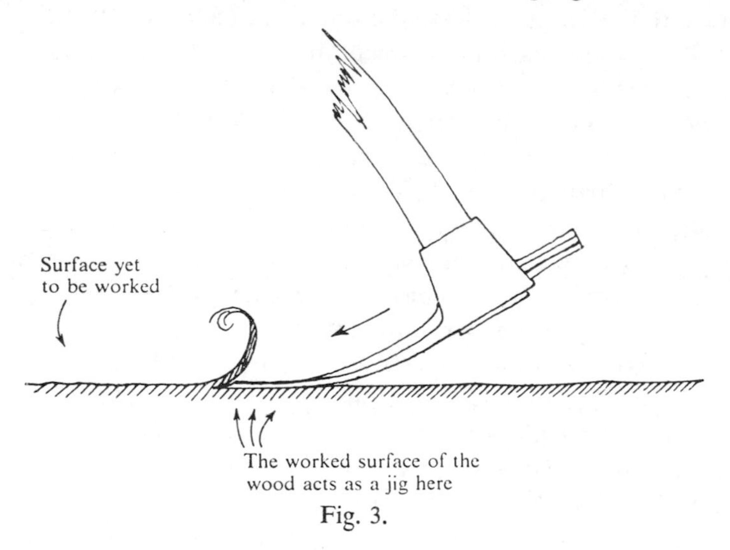

Fig. 3.

these in even one trade can be very large. In the first place many tools are partly self-jigging. The adze is, for one. The whole secret of using it accurately is that the curved back of the descending adze strikes tangentially on the flat surface left by the previous stroke—which becomes a partial jig—and rides along it so that the new stroke more or less continues the plane of its predecessor (fig. 3). In the second place, there are different degrees of certainty in jigging. Thus, if you want to cut a piece of notepaper straight, parallel and three inches wide, you can go to work in six different ways. Either (1) mark the line on the paper, take a knife, hold your breath, and run the knife along the line: in which case you are relying on dexterity; or (2) you can cut a little outside the line and then trim back to it by paring off many little narrow slips of paper in succession: in which case you are relying on gradualness. Or (3) you can cut along the line with scissors, which, like the adze, are partly self-jigging because in their case the newly cut edge of the paper butts against the upper blade of the scissors and steadies the sheet while they continue the cut. In this case it is easy to make a good job of the cutting. Or (4) you can cut with a knife along a ruler; the ruler is an effective jig and high regulation is still more certain than with the scissors. Or (5) you can use a guillotine, in which case it is really quite difficult to avoid high regulation, for the operation is now completely jigged. Or (6) the guillotine could be fitted with a fence and an automatic feed of some sort, in which case you would have the workmanship of full certainty and you could produce thousands of identical strips of paper.

Now, of these methods of paper cutting, 1, 2, and possibly 3, will show moderately free workmanship. 4, 5, and 6 equally will show high regulation, and short of actual failure in workmanship it will be impossible to tell the results of them apart.

In free workmanship the flat surface is not quite flat but, when seen from close by, shows a faint pattern of tool marks: and the straight edge is not quite straight, but, seen close, shows slight divagations. The effect of such approximations is to contribute very much to the aesthetic quality in workmanship which I shall call *diversity*—and which will be discussed in chapter 6. The natural figure of materials such as wood, the play of light in translucent materials, and the effects of wear, weathering and age, all contribute to diversity as well, but controlled freedom in workmanship has perhaps contributed more to the quality of our environment by way of diversification than any of them.

Free workmanship is now rare and becoming rarer. The workmanship of certainty is, simply of its nature, incapable of freedom. In old days free workmanship was the way of turning out cheap goods in quantity, but now even the smartest workman using it could not compete with the workmanship of certainty, and it survives successfully only in making a few things which the workmanship of certainty is incapable of, such as baskets and the products of the underwood industry, palings, spiles and hurdles, which are still in demand and have no acceptable substitute as yet. It is essentially deft, done with economy of effort. The liveliness and decision of it, and the fact that it is often associated with the countryside, have caused it to be thought that its practitioners must have taken pleasure in doing it: which may very much be doubted in some cases. The Welsh turner, James Davies of Abercych, told me that as a boy he had carved wooden spoons to be sold at fairs at, I think, twopence each. He said that at that price there was just time, when the spoon was finished, to look once at the inside, once at the outside, and then throw it over your shoulder on to the heap and start another! But having seen his work I do not doubt the spoons were a pleasure to look at.

Smiths are great exponents of free workmanship, for their trade above all needs deftness and decision. 'Strike while the iron is hot' is a very apt proverb.

There is no substitute for the aesthetic quality of this workmanship and the world will be poorer without it, particularly the countryside. It is impossible not to regret that it is declining but quite impossible to expect that it will survive on any scale as a means of decent livelihood.

It will be as well now to define the other main terms used in this book and to make explicit the relationship between them. I have introduced new terms with great reluctance and have tried to select words whose ordinary meaning would be violated as little as possible. I have, however, had to limit the meaning of a few common words, much as in scientific terminology 'stress' and 'strain' have very precise meanings which though derived from their ordinary ones are more circumscribed.

Before considering the definitions the reader may find it convenient to refer to plates 1–10*b*, and to the commentary on them in chapter 12, which is largely concerned with demonstrating the meaning of terms which have been introduced in the previous pages and with illustrating some of the points made in them.

Definitions and terminology are crucially important. A large part of the fruitfulness of scientific thought has come from one simple fact. It is that hitherto every scientific term has had an exact definition, verbal or mathematical, universally accepted. As a result communication in scientific terms between scientists has till recently been almost completely effective. Yet, on questions of art, communication is seldom so much as half effective. There is an immense amount of noise and little else. Definitions are the only possible basis for communication and we must have them. If they cannot yet be made final we must have provisional ones.

...Tzu-lu said, If the prince of Wei were waiting for you to come and administer his country for him, what would be your first measure? The Master [i.e. Confucius] said, It would certainly be to correct language. Tzu-lu said, Can I have heard you aright? Surely what you say has nothing to do with the matter. Why should language be corrected? The Master said, Yu! How boorish you are! A gentleman, when things he does not understand are mentioned, should maintain an attitude of reserve. If language is incorrect, then what is said does not concord with what was meant; and if what is said does not concord with what was meant, what is to be done cannot be effected...*

In the present context the important point when defining *design* is to distinguish between it and workmanship, and for this present purpose we may define design as whatever can be conveyed to the workman by drawings and by specification in words or numbers. For a fuller definition of design the reader is referred to *The Nature of Design*. In that book design is differentiated from invention and the essential nature of the activity of designing is examined.

The *designer*, as the term is used in the present context, means a person or group of people who decide the contents of the drawings and specification: that is to say, decide what information they are to convey. (The designer may of course also be the maker.)

The *intended design* of any particular thing is what the designer has seen in his mind's eye: the ideally perfect and therefore unattainable embodiment of his intention. The design which can be communicated—the design on paper, in other words—obviously falls far short of expressing the designer's full intention, just as in music the score is a necessarily imperfect indication of what the composer has imaginatively heard. The designer gives to the workman the design on paper, and the workman has to interpret it. If he is good he may well produce something very near the designer's intention. If the workman is himself the designer he almost certainly will (but that does not imply that the designs a workman intends are necessarily good ones).

Now it is by reference to the intended design that we judge the quality of workmanship, and we have to infer the intention from what the workman has done. Moreover, the intended design will have been conceived in terms of the kind of workmanship, that is to say, the degree of regulation, which is economically suited to the product. Thus it is not possible to judge the quality of workmanship unless we have prior knowledge of that, and unless also we are in a position to judge what the designer might reasonably have been expected to intend. In times when there are established traditions in all

* *The Analects of Confucius*, Book XIII, 3 (translated by Arthur Waley, 1938).

branches of design and workmanship any moderately cultivated man will fulfil these requirements and will be a good judge of workmanship; but when, as now, traditions of design and workmanship are in flux, rapidly changing in many fields, more discrimination is needed.

The intended design of a thing and the *ideal form* of it may be, but are not necessarily, two quite different things. The ideal form is the most highly regulated form, and more highly regulated still than that. It is conceivable but not attainable: the perfect cylinder, the perfect rectangular prism, the perfect sphere. But the intended design so far from being concerned with these perfections may perfectly well be concerned with rough-hewn billets or cleft oak rails (plate 6). On the other hand the modern engineer's intended design, and the architect's, very often are conceived in terms of ideal forms: flat planes, straight edges, perfect cylinders, arcs of circles. This can happen because these are the forms which can be communicated with least trouble. Fully to describe the form of the billet in plate 6 would take any draughtsman a matter of weeks.

Since the quality of workmanship is judged by reference to the intended design, it follows, as everybody knows, that what is good workmanship in one context is bad in another. The workmanship is good in plate 1 and also in plate 4, but if parts of the cabinet were finished like the carving it would be a sorry affair indeed. It is possible in the same piece of work to put high regulation and quite free workmanship side by side, but unless the two are evidently dissociated (plate 25) very nice judgement is needed. Usually the highly regulated parts make the free parts look careless.

Technique is the knowledge of how to make devices and other things out of raw materials. Technique is the knowledge which informs the activity of workmanship. It is what can be written about the methods of workmanship.

Technology is the scientific study and extension of technique. In ordinary usage the word is slapped about anyhow and used to cover not only this, but invention, design and workmanship as well.

Workmanship is the application of technique to making, by the exercise of care, judgement, and dexterity. As opposed to design, workmanship is what for practical purposes the designer cannot give effective instructions about by drawings or words, although he can envisage it perfectly well. The designer is apt to imagine he has more control over workmanship than he has.

Standards of workmanship become established in each kind of manufacture. The designer gets used to them, expects them, asks for them, and comes finally to believe he is getting them because he asks for them. Then he comes up against a firm who do not know their work and finds he is helpless. All he can do is to say 'do it again'. When the work is bad the second time, his resources are at an end. You cannot compel or command good workmanship by the terms of a contract.

Suppose that the designer can feel entirely confident that any of twenty or thirty firms working to the same drawings will turn out a nearly identical job, that is no evidence at all that the same drawings have actually enforced the same quality in each case. It simply means the firms are using about the same technique and are working to the same standard, being competitors; no matter who sends the drawings or what is in them.

Good workmanship is that which carries out or improves on the design, whether the design was made by the workman or another.

Bad workmanship fails to do so and thwarts the designer's intention in respect either of soundness or of comeliness: or else it makes things look equivocal in the sense that the material used appears to have, simultaneously, properties such as hardness and softness, or characteristics such as roughness and smoothness, which are incompatible with each other. A kind of equivocality is also produced by putting together formal elements, features, which are incongruous, such as a polished surface with a raw or jagged edge (chapter 9).

Skill is a word not used in this book. It does not assist useful thought because it means something different in each different kind of work. To a smith, dexterity is important but rarely in the extreme; but his judgement of certain matters, particularly heat, has to be brought to a pitch and decisiveness rarely needed or matched in woodworking trades, in which, however, more dexterity is often needed. Moreover, much of what is ordinarily called skill is simply knowledge, part of 'what can be conveyed by words or drawings'.

There is an old saying that when you have learnt one trade you have learnt them all. There is truth in it. Beside the special forms of dexterity and judgement which belong to any one trade, something general is learnt which makes it easier to learn others, though still not easy. This may be merely the habit of taking care but it seems to be more.

At all events 'skill' is ordinarily used to refer to an uncertainly distributed group of disparate things. Like 'function' you can make it mean what you please. It is a thought-preventer.

In the *workmanship of certainty* the result of every operation during production has been predetermined and is outside the control of the operative once production starts.

In the *workmanship of risk* the result of every operation during production is determined by the workman as he works and its outcome depends wholly or largely on his care, judgement and dexterity.

There are nowadays two quite distinct purposes to which the workmanship of risk can be applied. One is *preparatory*, the other *productive*. Preparatory workmanship makes not the products of manufacture but the plant, tools, jigs, and other apparatus which make the workmanship of certainty possible. The workmanship of risk should, I suggest, be termed 'productive' when it is used actually to turn out a product for sale.

The workmanship of certainty is almost invariably regulated and the workmanship of risk often is also. *Regulated workmanship* means workmanship where the achievement appears to correspond exactly with the idea; things meant to repeat appear to repeat exactly, things meant to be square look exactly square, and so on. If, on the other hand, they do not appear to repeat exactly or to be exactly square, then there is an evident disparity between idea and achievement, there is approximation, and we call the workmanship *free*, or, in cases where the disparity is very large, *rough*.

In the workmanship of risk, in all trades, the course which historical development has taken has usually been to increase the workman's power to regulate, and the standard of regulation aimed at has tended to get higher. There are indeed many instances where the workmanship of risk achieves higher regulation than that of certainty: for instance in the production of an accurately flat surface on a machined casting, or in optical work.

The workman is the term I propose for a man, woman or group of people who interpret and execute a design by the workmanship of risk using judgement, care and dexterity. I propose to use it as a generic term, like 'the executive' or 'the judiciary'. The workman is thus the agency which provides the tools, jigs, prototypes and other material basis for mass-production. In most trades the workman will at one moment be working freehand while at the next he will be applying the workmanship of certainty, making use of jigs and machine tools; and the preparation and combination of a series even of completely jigged operations or machining operations always involves judgement and care if not invariably dexterity.

By an oddity of usage 'workmanlike' is a laudatory word and 'workmanship' is at least neutral, while the word 'workman' tends to be used as though it meant the same thing as 'labourer'. I have a respect for many labourers, but I do not intend that meaning. One can no longer use the word 'craftsman'. Its tendency is unsure. It is becoming rather a word associated with wage claims or now with artist-craftsmen only. To call a man a good workman should imply the highest respect, just as it once did to call a man a good seaman, no matter whether he was a shipmaster or an A.B.

It was the workman of other days who made, by the workmanship of risk, the tools, jigs and prototypes which enabled the machine-tools of our day to be built and with which in turn the workman of our day is now making the material basis for automation. The tools, jigs and machines on which the workmanship of certainty will always depend are simply the stored embodiment of the care, judgement and dexterity exercised by the workman at an earlier time. And if machine-tools are now able to breed other machine-tools it is only because the workman is their matchmaker, their midwife, their nurse, doctor, educator and much more.

In the Science Museum in London can be seen the first of all lead-screws, which Maudslay chased for the first screw-cutting lathe, and one of the first planers, whose bed Roberts chiselled and filed flat. How many generations of screws and plane surfaces can those two machines have bred?

'The workman' covers Stradivarius and it covers the monk who drew the Chi-Rho page in the *Book of Kells*: for the workman may or may not also be a designer.

In the workmanship of risk, decisions are very often made by the workman which could have been made by the designer, and the workman may himself be the designer. Consequently the term 'workmanship' is often used far more loosely than the definition I have proposed will allow. By that definition nothing is workmanship which a designer could alter by speaking or writing a word or two; and workmanship is the exercise of care plus judgement plus dexterity. These can be taught, but never simply by words. Example and practice are essential as well. It is no part of a designer's job to teach them, even though he may be able to.

By numerical control certain designs can be translated (not interpreted) and 'told' directly to a machine tool so that a prototype or tool can be made without any care, judgement or dexterity being exercised at this stage. Ultimately automation may dispense with the operative altogether; but hardly with the workman, who will presumably remain indispensable to it somewhere, even if numerical control advances to the point that a set of machines, given a suitable programme, can design and make another without the workman intervening at all.

'The workman' is to stand for a group of people just as much as for one. A group of people executing a design are closely analogous to an orchestra and decidedly not to a team. A team has either a driver with a whip, or another team opposing it. In an orchestra each player—(workman)—is interpreting—(working to)—the same score—(design)—and is called on to play the instrument—(apply the technique)—in which he is expert, at the stage in the performance where it is needed.

The workman is essentially an interpreter. It is usual to suppose that sometimes, in Ruskin's words '...the thoughts of one man can be carried out by the labour of others' because the design is 'determinable by line and rule'. If the designer, so-called, has no interest in the appearance of the job, his thoughts will be so crude that this may even be true, in fairly simple cases. But, where, on the contrary, the designer is a responsible man, it can never be true. It is no more possible for an Act of Parliament to determine the law than it is for a 'design'—meaning an affair of drawings and instructions—to determine the appearance of a thing. There are judges to determine what the Act means after Parliament has done its best to make its intentions clear. It is for the workman to determine what the designer means after he had done his best. So it is for the conductor or pianist to determine what Bach means after he has done his best, by means of a score. The judge, the pianist, and the workman are interpreters. Interpreters are always necessary because instructions are always incomplete: one of the prime facts of human behaviour.

No drawing, however fully and minutely dimensioned, can ever be more than a sketch as regards the appearance of the thing drawn. The eye and mind discriminate things which can never be specified or dimensioned: the qualities and colours of surfaces, the minute variations of profiles, and still other nuances of shape too tenuous and subtle to describe in practice.

John Dreyfus's book, *Italic Quartet*, is a remarkable account of an undertaking defeated by a very good workman's inability effectively to interpret a kind of design to which he had not been brought up. It shows that Edward Johnston, the great calligrapher, clearly understood—and his patron in this case did not—that, workmanship being interpretation, the quality of a type in his time ultimately depended not on the designer but on the punch-cutter; for all the meticulous instructions that a designer might give him, and that Johnston in this case did give him, and which are illustrated in the book.

Although workmanship is interpretation, I do not suggest that anything a workman gives us can be as moving as what the performer of music does. No design lives in the same world as music, and the performance of music allows more subtle and deep expression than any workmanship can possibly attain to.

5

The designer's power to communicate his intentions

The definitions of design and workmanship proposed in the last chapter raise the question: how far is it possible for words, figures and drawings to prescribe the qualities of a work of art; so that if the designer's directions are faithfully obeyed the qualities automatically arrive? Is it really necessary that anything should be left to the workman's discretion?

On what properties of matter do the visual arts depend? From which objective, defined and measurable properties of things do we derive subjectively those indefinable qualities which are the stuff of visual art? Clearly there are very many physical properties, such as thermal conductivity, which do not concern us at all. Only size, shape, reflectance, colour, and translucency are important here. Each one of these is measurable, shape as much as the rest; for any shape can be defined by co-ordinates of points on its surface as in a graph, and if the intervals between the co-ordinates are made very small the definition can, in principle, be made virtually complete. But co-ordinates, of course, will not always be necessary. The simple geometrical shapes characteristic of architecture and of mechanisms and other constructed things are so easy to define that even the familiar crude dimensioned diagrams which designers call drawings are sufficient.

Thus, in principle, it is possible for a designer to prescribe quantitively all the properties by means of which any given object is judged to have the qualities of a work of art; and to give an absolutely complete description of it. In principle nothing whatever is beyond the reach of design. In principle it is possible to define all the properties of the crock-lid in plate 7 so completely that a manufacturer who had never seen it or even a photograph of it could produce fifty thousand perfectly indistinguishable replicas of it in plastic. And in principle, no doubt, the chemistry of claret from *this* vineyard in *that* year can be analysed so completely that the wine can be perfectly reproduced in any desired quantity. In practice things are different.

There is nothing abstruse about practicability. These things are impracticable in design simply because no one would pay for the immense amount of exacting work which would be involved, first on the designer's part in making complete definitions, and then on the workman's in trying to comply with directions so voluminous as to slow his normal pace almost to a standstill. The cost of designing for quantity-production is high in any case, and so there is a strong incentive to design only in terms of shapes which are easy to communicate: either those of standardized components, or those geometrical shapes which are easily and automatically formed by the standard and readily available machine-tools on which so much of the workmanship of certainty depends.

This impoverishment is the price we pay at present for cheap quantity-production in which only this very simplified level of communication and execution is practicable, and in which as a rule the slight free modifications of shape and surface quality which mark the workmanship of risk are quite unattainable and indeed unthinkable, except in cases where the material is flexible or translucent (for the latter, see plate 30).

At this point the reader may find it convenient to refer to plates 22 *b*–31 *b* and the commentary on them, which mainly deals with the difference between good and bad workmanship and the question of the designer's intention.

6

The natural order reflected in the work of man

In every natural organism we see a dichotomy between idiosyncrasy and conformity to the pattern of the species. No two leaves of the same tree are precisely alike, each is individual: yet every one of them conforms to a recognizable pattern characteristic of the species. No oak looks like an ash. So it is with human faces, each individual, yet each recognizably conforming to the human pattern, the idea or norm of the human face. There are no cyclopses.

In this we see an obvious parallel with the disparity between idea and achievement in free workmanship. There also we recognize the underlying idea, the ideal form—the 'pattern of the species'—and yet we see it given individual expression because of approximate workmanship. Thus in free workmanship we see the natural order reflected in the works of man.

In nature we see varying degrees of disparity between the idea and the achievement wherever we look. To Plato it may perhaps have seemed that things would be better if there were no such disparities. We, having lived in an age where to all appearance such disparities really can be banished from our environment, may doubt it. There have been men for many thousands of generations, for tens of thousands. Only the last couple of hundred or less have started to cut themselves off from their natural environment: too few generations for any significant evolutionary change. If the appearance of the environment matters a little to us all, as to some of us it matters overwhelmingly, then it seems reasonable to suppose it may be good for us to import into the unnatural environment we have made some of the quality of unmonotonous unexpectedness that our race was born to live with.

Our traditional ideas of workmanship originated along with our ideas of law in a time when people were few and the things they made were few also. For age after age the evidence of man's work showed insignificantly on the huge background of unmodified nature. There was then no thought of distinguishing between works of art and other works, for works and art were synonymous. All, however crude, were more or less admirable simply because they were rare, and of immense importance to their users.

Then and for long afterwards—and even now in some remote places—all the things in common use for everyday purposes were of fairly free or rough workmanship and anything precise and regular must have been a marvel, amazing and worshipful.

So it came about that, as soon as civilization emerged, with specialization of labour and professional craftsmen, they strove for precision and regularity and turned against rough workmanship. Precision and regularity symbolize mastery. The Pyramids are a witness that unadorned precision alone will convey majesty if the scale is large enough.

This reverence for precision had, I think, two explanations. The first but less compelling of the two is that precise, regular workmanship of any importance necessarily implied specialization of labour, and that in turn meant that if you were able to pay for it you must be rich. It was immediately a proof of your personal power.

The second, and I believe deeper, reason lay in the opposition of art to nature. The natural world can seem beautiful and friendly only when you are stronger than it, and no longer compelled with incessant labour to wring your livelihood out of it. If you are, you will be in awe of it and will propitiate it; but you will find great consolation in things which speak only and specifically of man and exclude nature. When you turn to them you will have the feeling a sailor has when he goes below at the end of his watch, having seen all the nature he wants for quite a while.

Precision and regularity in those days signified that, to the extent of his intellect, man stood apart from nature, and had a power of his own.

It is really very difficult indeed for us to realize what precision and regularity must have meant and how moving they must have been, when now they are seen in every trivial product of a money-bound society: in the throw-away ball-point pen and the tomato-can. The reflections of the natural order which we see and value in rough or free workmanship must then have seemed far less amiable, a mere reminder of what men wished to part from and be less involved with, now that they lived in cities. We, on the other hand, would do better to make things occasionally so that they do reflect our community with the natural order instead of emphasizing our separation from it; and so that their diversity would stave off the monotony which comes of too much regularity and precision. What was their meat will soon be our poison.

Until the general advent of the workmanship of certainty, high regulation used to be, to some extent, a mark of honour or respect. Anything fit for a king must unquestionably be of highly regulated workmanship. One sometimes finds that, when a common thing has acquired a symbolic value in some particular context, a curious highly regulated version of the rough workmanship which rightly belongs to it is done. The Woolsack may be stuffed with wool but it does not look so much of a sack as other sacks do.

The contrasting qualities of workmanship, precision and approximation, regulation and freedom, are neither good nor bad in themselves. There are kinds of precise work which are unprepossessing and others which are exquisite. The same can be said perhaps even more strongly of free work. A brisk element of improvisation reflecting the natural order is one thing; but there is an element of improvisation in plain bad workmanship too. The use of free workmanship no more guarantees aesthetic quality than does the use of oil-paint. But even supposing that every bit of regulated and of free work were good of its kind there could still be no question of establishing the absolute superiority of one kind to the other. Their value is relative to their time and circumstances. Regulation once had a meaning which it no longer has; while free workmanship begins to mean what it can never have meant before.

7

Diversity

We hear music by virtue of the relationships in sound between different notes and their relationships in time. We see art, which includes design, by virtue of the relationships in appearance between the different visible features of a thing, its formal elements, and by virtue of their relationships in space. The point I wish to make is that design—the music of design—depends on the relationships between distinguishable and separable features of things which are to a certain extent analogous with the elements of music, its notes and chords.

The variety of features or formal elements is infinite. The designed shape of a column, the surface quality of it, its colour seen at a certain distance in a certain light, the pattern in its material, the joints in it, the lichen or dirt which time has given it—all these features we may term formal elements, though not all of them were designed. An element may be obvious or barely perceptible. Entasis in the profile of a column might be very slight yet still very important aesthetically. Very slight deviations from a regular profile due to approximating workmanship would also be important elements.

It is a matter of the greatest moment in the arts of design and workmanship that *every formal element has a maximum and minimum effective range.* It can only be 'read'—perceived for what it is—by an observer stationed within those limits. The slight deviations from regularity in the profile of our column will have become imperceptible, probably, by the time your eye is four feet away from it; but the cylindrical designed form of it will still be clearly perceptible at several hundred yards. Yet again, if we look at the column through a magnifying glass, its cylindrical form will be imperceptible because we are viewing it at less than the minimum effective range for that element.

Every little incident of form and surface and every departure from regularity however minute will begin to tell as a formal element at some particular range. Of course, if its maximum range is so short that a magnifying glass is needed, then for our purposes the element is negligible. So may the great masses of a building be negligible, if you have not opened the range enough to enable you to see them in relation to each other. You cannot see the building from the doorstep.

In nature, and in all good design, the diversity in scale of the formal elements is such that at any range, in any light, some elements are on or very near the threshold of visibility: or one should say, more exactly, of distinguishability as elements. As the observer approaches the object, new elements, previously indistinguishable, successively appear and come into play aesthetically. Equally, and inevitably, the larger elements drop out and become ineffective as you approach. But new incidents appear at every step until finally your eye gets too close to be focused. The elements that at any given range, long or short, are just at the threshold, that we can just begin to read, though indistinctly, are of great importance aesthetically. They are perhaps analogous to the overtones of notes. They are a vitalizing element in the visible scene. They are indeed found in every natural scene except for those which most depress us, like a white wall of fog, or an evenly overcast whitey-grey sky. But they are not always found in the environment man has made for himself, though formerly they always were. That explains the blankness we often find, now, in the expression of a product or a building when we get close up to it. Down to a certain distance everything about them looks well. As you close the range after that point nothing new appears. There are no further incidents. As soon as you get towards the minimum effective range of the larger, designed elements, the whole thing goes empty.

As we have already remarked, workmanship provides formal elements, and important ones, which are outside the control of design: of what, for practical purposes, can be conveyed by words or drawing. These are, of course, short-range elements. Most of them are still at, or little above, the threshold of recognition at those close ranges at which we normally see the components of our environment when we are using them: in a room, in a vehicle, in a street, on a bench or table, in our hand. For most of your life the parts of your environment which you are looking at are likely to be at close ranges of that sort; not on a hilltop, or in the distance, or as seen in the photographs in architectural magazines. It is for this reason that the art of workmanship is so evidently important. It takes over where design stops: and *design begins to fail to control the appearance of the environment at just those ranges at which the environment most impinges on us.*

A thing properly designed and made, continually reveals new complexes of newly perceived formal elements the nearer you get to it. Any considerable building will reveal itself differently at every range from six inches to several miles. A rubbish heap also will continually reveal new formal elements as you approach it, and of the most diverse sorts, but since there are no ordered relationships between them there is no quality of art about it.

Now, many of the formal elements revealed on a close approach to anything, even if it is of the finest workmanship, are commonplace; but then most formal elements in themselves are commonplace. It is in the relating of them to each other, and often in the subtlety of those relations, that the art lies. In general, of course, there is far less scope for new formal invention in workmanship than in design, because the possible ways of relating the familiar formal elements which recur in workmanship are often few. The scope varies according to the technique. But, if there is little scope for innovation, that does not impugn the importance of workmanship to art. Art has nothing to do with the fact of new invention, it resides in the quality of what has been invented; and whether the invention was made recently or not is irrelevant to the standing of the work of art. We do not burn our Rembrandts. Novelty can be exciting and delightful in art as in other affairs, but art exists in its own right, independently of novelty.

In the art of workmanship, then, we seek to diversify the scale of those formal elements which begin to be distinguishable at close range and also—in season—to diversify the forms themselves by allowing slight improvisations, divagations and irregularities so that we are continually presented with fresh and unexpected incidents of form.

It is rarely possible to do this by the workmanship of certainty, but always possible by the workmanship of risk, and particularly easy by free workmanship. There is a very present danger that, as the kinds of medium-scale diversity which free workmanship used to impart to building become less readily available, what little can be had in that way will be over-played and in the end travestied. We do not want every piece of concrete to show board-marks, every piece of paving to be cobbled, every piece of masonry to be random rubble, every piece of brickwork to be left unplastered. There is a place for all those things, but such elephantine capers unaccompanied by diversity at smaller scales become merely ludicrous. What we want is diversity which begins at the smallest visible scale and develops continuously upwards from that; and even then we do not want it always and everywhere. Vitamins are necessary to life, but only in small amounts. Take them in large amounts and they make you ill. So I believe it is with diversification in workmanship. I do not suggest it is more—or less—than a vitamin: not a diet: not a panacea: merely something which, though we may not take much notice of it, we need to have.

Nor am I saying that free workmanship is better than regulated, nor that regulated workmanship is the ruin of our civilization. On the contrary, I say that on the contrast and tension between regulation and diversity depends half the art of workmanship. But for our generation unrelieved regulation is bad, and may even be dangerous.

Some contrast and tension between regulation and freedom, uniformity and diversity, is essential, and it is the play made with it which most sharply characterizes human as distinct from natural workmanship. The delight which has always been felt in things made of wood and marble rests mainly on the contrast between the regularity of their design and the diversity of the material. If, however, you cover a large flat wall with unrelieved wood veneer, even though of excellent quality, the effect is vapid; for the figure of the wood destroys, in appearance, the flatness of the wall and there is no element of contrast left except at the corners. Moreover, all the formal elements are in the figure of the wood, all of one character and having little variation in scale.

Yet other examples of contrast are found in machinery. There, in a painted casting, you have the rough sandy diversified surface set off against the absolute uniformity of the colour on it; and there are many other contrasts such as that between the absolute uniformity of a polished cylindrical part with sharp highlights, and the very slight diversification by light and shade of a turned rod with a tool finish on it: or between the 'frosted', highly diversified appearance of a scraped surface and the regularity of a planed one; or, at very close range, between the minute regularity of a machine-tool finish on a flat part and the controlled freedom of another part finished by filing.

The effects of age and wear are powerful diversifying agents, and it is appropriate to consider them here. As every workman knows, they begin to leave their mark on a thing even before its making is finished. On good workmanship their effect is often beneficial, and it seems a little less than obvious why we are so fond of new things. Perhaps, then, we had better consider that point before we proceed.

When and why do we prefer things not to have been affected by age and wear? There are in the first place arbitrary reasons of prestige. To be smart, one should have a new car and an old house: or perhaps that is now out of date and one should have an old car and a new house: in either case these whims of oneupmanship need not detain us. A more important and constant reason may be that we do not like to think of ourselves ageing, and project this feeling on to our possessions. When we renew them we half imagine we are renewing ourselves.

It is probably true in general that we like things new because of what newness symbolizes rather than for any special aesthetic qualities inherent in it. There are, however, obvious exceptions, such as things made only for a short life, which quickly lose their youth and look worn out: which, in other words, soon get damaged. Damage is the name we give to any kind of accidental change which thwarts design, in respect of either soundness or comeliness. When the damage happens to be done by mistake during the process of making and is then repaired, as often happened in rough work done by inexpert men, the repair is called a botch.

By no means all accidental change is damage, and much of what is not is capable of improving the look of things. Patina and distressed surfaces of one sort and another have been prized from ancient times. So have such accidental or originally accidental modifications as crackle glazes, weathering of stone, fading of wood, moiré silk, and boarded leather. Some will argue that the soot on St Paul's Cathedral used to enhance it. Not I! But overgrowths of lichen and some other sorts of dirt may be most beautiful.

All these amount to adventitious formal elements. If they do not thwart but rather amplify the design by introducing elements of diversity which it lacked, then we like them. If they seem interesting in themselves but yet remain extraneous to the design—as I think the soot on St Paul's did; for it seemed to me to act like disruptive camouflage—then we are apt to dislike them.

Some kinds of workmanship rely for their effect almost entirely on the adventitious diversity introduced by certain materials, of which wood is the prime example. Wood can be selected and cut so as to control its figure, but only to a limited extent; and indeed no one wants to control it too much.

The beauty of cabinet-work is in the infinite diversity of the wood setting off the precise regulation of the work.

In our society at present the sensitivity of people to the quality of diversity in workmanship seems very uneven. Almost anybody will at least pay lip-service to the qualities which age and weather impart to the outside of buildings; though it is often a pretty naïve tribute compared with the Japanese cult of *sabi*—'the love of imperfection as a measure of perfection'.* But the qualities imparted by wear to good furniture are often clotted over with polish of one kind or another. Similarly, while new furniture still has to look as though there were wood in it, the properties of the wood grain as a diversifying agent at short range are often nearly extinguished by stain and lacquer.

It is probable that the insensitivity to diversification which is apparent in so much industrial design is partly the consequence of the shortcomings of photographic reproduction. It is said, truly, that painters have formed our vision of the world; but what they have done in that way is as nothing compared with what the photographers have imposed on us, though unintentionally. Things are designed with future photographs of them in mind. Good photographers—who are not common—can show much, though never all, of the diversification, but in any ordinary cheap reproduction most of it disappears and even with the very best possible reproduction a good deal is lost. Sometimes one suspects that what the eye does not see in the magazine illustration the heart of the designer does not grieve for. At least such illustrations do very little injustice to some of the designs they show, while anything of reasonable quality shown in the same way would have been emasculated; for the things which are caricatured or degraded in those cases are chiefly the things which do stand, and need, examination at close range: the range at which they are mostly seen in use, though not in photographs.

A good reproduction of a good photograph can show an astonishing amount, to be sure, and it rests on considerable feats of workmanship; but it always tells less than the truth.

Yet we are all aware of diversification. Painted marble or a waxwork figure look as good as real at a distance, but seen close-by they are insipid because their diversity falls short of that in real marble and real skin. But then the formal elements revealed at a close view of marble, or wood, or fine

* De Bary, *Sources of Japanese Tradition.*

workmanship, are very subtly differentiated, while in this age quantity is what we like; something that shows up well: for 'big' and 'important' are about synonymous in our conception of art.

After so much about diversification in workmanship it had better be repeated that diversification is not essentially a property of workmanship alone, but that at medium and long ranges it is entirely controlled by design, and at long range usually with great success. We take this for granted, indeed, without ever thinking that it is a fact connected with the theory of workmanship. Distance lends enchantment. An ugly building is apt to look less ugly the farther off you go. This happens not only because you can no longer see the ugly elements, but also because the elements you can see, the main masses of the building which are still effective as formal elements at long range, often make a fairly successful design, and perhaps still more because the details which were ugly while still in range and recognizable make a satisfactory undercurrent to the whole effect when they are only at the threshold of recognition.

The sight of a good building or a ship at a long distance in very clear air has a particular attraction for people in Britain because the effective range for recognition of many features of it will have been increased by the weather. Thus we see in distant miniature the whole gamut of diversification which British people can normally see only in the middle distance because of their slightly misted air.

It may not have been suggested before that the downward extension of design to the minutest scale of workmanship is governed by the same law which determines the appearance of a distant mountain or gigantic building, or yet that the elements on the threshold of recognition are important at every range.

Ruskin knew and wrote about diversity, though he did not call it by that name. Characteristically, he observed the facts with great insight and discrimination and stated them beautifully; but drew conclusions from them which they do not point to, such as 'The art of architectural design is therefore, first . . . the rejection of all the delicate passages as worse than useless and the fixing of the thought upon the arrangement of the features which will remain visible far away.'* Why, in that case, he found the delicate passages in nature so moving, it is not easy to understand.

* *Stones of Venice*, chapter XXI, para. XXI.

The aesthetic importance of the downward extension of diversity has long been recognized in the Far East, but in the West imperfectly and intermittently.

At this point the reader may find it convenient to refer to plates 10*c* to 22*a* and the commentary on them, which mainly deal with the subject of diversity.

8

Durability

Time is a dimension of all workmanship. It all fails, to be sure: but it fails either sooner or later. Durability is thus a preoccupation of every workman.

At any given time each trade has accepted standards of what are good methods, and methods are often reckoned good solely because they are durable. Beyond that, whenever a method can plausibly be said to make for durability, then it is said to. Yet often enough one finds examples of 'bad' methods or workmanship which have lasted just as long as the good ones (plate 23*b*). It is impossible to resist the conclusion that many of the 'good' methods have been preferred, in reality, for aesthetic reasons: and of course quite a number of the preferences are avowedly aesthetic. No one supposes that secret mitre-dovetails are any more durable than lapped ones.

As for workmanship, as distinct from technical method, it is questionable whether the high regulation so often considered essential to a 'good job' where the workmanship of risk is in use, really makes for any more durability than rough work would do, in many cases. Part of the high regulation done by the workmanship of risk has always been simply an affair of art, of doing a thing in style, and of no use whatever (but though of no use it is of great value as an ingredient in the quality of the environment). In a machine or instrument, high regulation is useful only in cases where two parts slide on, or in, one another; or in a valve seat, or a bearing, or where a tube or spring of constant section is needed; but the rest of the high regulation on the visible parts of the components is useless apart from these cases and others similar to them. In any machine we can find things which are straight or flat but have no need to be, or which are smooth, or polished, or cylindrical, or precisely similar to each other, but would work just as well if they were not. So also it is with buildings. They can be wonderfully crooked and lopsided but no less useful for it.*

With the workmanship of certainty to supply standard components, it is

* C.f. *The Nature and Aesthetics of Design*, chapter 9.

usually cheaper and easier to produce high regulation than to do without it. But these things used to be done on all hands by the workmanship of risk, and with that they were by no means cheap to do. There really seems to have been a tacit conspiracy between designers and workmen to suggest that more of this high regulation is necessary than is the case. They meant to do it for the sake of art even if their clients did not always fancy paying for it. But to make it a fair deal they gave the client his money's worth by making sure that, even if the job cost more than in strict necessity it had to, still it would last a very long time.

Durability, in fact, does not require as much high regulation as is often thought, but has been used pretty freely as an excuse for doing highly regulated work; so much so, that high regulation has usually been reckoned a sure sign of durability: a mark of 'the best quality' which one buys because it will last. But a very rough job may last just as well and so, it must be repeated, may a bad one. I remember an Amati violin—an authentic one—which I was shown after it had been opened for repair. It was a shocking job! There were glue-lines thicker than one's thumb-nail, if not worse. Yet it had been singing its song to generation after generation and been treasured by them all because of it. It was a very good and durable violin in fact, regulation or no regulation, but not by any means up to later standards of workmanship.

The traditional association between high regulation and durability whether true or false has obviously no force any longer. The highly regulated ball-point pen with which I am writing will be thrown away next week.

We have already remarked that traditional ideas of workmanship originated when man-made things were few and highly prized, of whatever sort they were, and when highly regulated workmanship must have been so rare as to seem wonderful. But now things are all too many, high regulation is commonplace, and free workmanship is rare. Thus both the old respect for workmanship as such is fast dying out, and high regulation, of all things, is least respected. Consider any scrapheap.

Ruskin said 'If we build, let us think that we build forever'. Shall we say 'If we build, let us remember to build for the scrapheap'? Shall we make everything so that it goes wrong or breaks pretty quickly? I think not. Men do not live by economics alone. There is a question of morale involved. A world in which everything was ephemeral would not be worth working for.

There are overwhelming social and aesthetic arguments for durability in certain things even if, as we are told, there are no economic ones. These are:

First of all, the things we inherit from the past remind us that the men who made them were like us and give us a tangible link with them. This is a thought to set off against the knowledge that life is short. Hitherto it has been inconceivable that any one generation should discard all the equipment it has inherited and replace it completely. That may yet become possible. Even if it does, it will still be imperative for each generation deliberately to make some of its equipment so that it lasts and survives its makers.

Secondly, if you are making a thing so that it goes wrong or breaks, then, however honestly you state the fact, two other facts remain. One is that you are putting as little into the job as you decently can. The other is that you are in a fair way to force its user to spend his money on replacing that thing instead of for some other purpose. He may be glad to replace it, in an age of materialism and the passion for novelty. But why should we all be compelled to keep spending money on renewing our car, our cooker and our refrigerator? These things for some people are merely means to other ends in life. Why should we not save the money so as to pursue those ends the better: altruistic, learned or artistic ends, say? Things which are made to fail early should be made maintainable and repairable, so that a man who cares for something other than novelty and status-symbols can make them last his time respectably while he gets on with his life. Optional durability is what we want.

Yet another reason is that age and wear diversify the surfaces of things in ways that nothing else will. If nothing ever lasted we should be denied that beauty. And yet another reason is that, where everything is ephemeral, novelty comes to be overvalued and mistaken for art; so that design is reduced to fashion pure and simple. There is an element of fashion in most design and one would be sorry if there were not, but it is only when a design has survived long enough to go out of fashion and be forgotten for a time that it can be appreciated or rejected for what is really in it.

By the definitions we have adopted, the durability of a made thing depends partly, perhaps largely, on workmanship where the workmanship of risk is used, but depends on design almost entirely in the workmanship of certainty: for there nearly everything which affects durability has been predetermined and can be specified by the designer. In some trades which use imperfect forms of the workmanship of certainty, however, it is customary for the workman to select material from a stock which the designer will have specified, and the workman may thus influence durability. This happens in chair-making and cabinet-making factories. It is, however, simply a matter of convenience and trade custom, not of principle, for in nearly all such cases the choice could be made by the designer rather than the workman. Similarly a deficient surface protection, which is a common cause of things being prematurely thrown away, because their appearance becomes unbearable, results from design also, as a rule. The designer will have specified how the surface shall have been prepared and what materials shall be applied, in what way, to protect it.

Thus premature failure, nowadays, can less often be blamed strictly on workmanship than is usually thought. It is only in the workmanship of risk, not that of certainty, that the workman's responsibility for durability is likely to exceed the designer's.

In the workmanship of risk the workman has seldom known for certain, on the basis of measured tests, which of the alternative methods available to him is the most lasting; and he has often over-estimated or over-stated the importance to durability of high regulation. But his preoccupation with durability has been a very real one. All the world knows that any good workman feels a responsibility for the durability of what he makes and feels bound at the very least to make the unseen parts of the job as sound as those which are visible; but his concern goes farther than that. Durability is apt to become for him an end in itself quite apart from moral considerations. He is apt to hold that a thing is not made properly unless it is made to last. That belief may be arguable now, but we have no cause to regret that it was acted on in the past and not much cause to fear it will be over-acted on in the future.

9

Equivocality

In addition to the bad workmanship which comes of thwarting the design, there is another kind which has the effect of suggesting that whichever material is being used has simultaneously a pair, or set, of properties such as hardness and softness, or objective characteristics such as roughness and smoothness, which are necessarily incompatible with one another. Yet another kind of bad workmanship having a related result consists in getting out of one piece of material adjacent formal elements which are incongruous, such as a polished surface and a raw edge. I propose to call the upshot of these defects *equivocality* and to discuss them in detail in this chapter.

The inconsistent properties and characteristics which produce the effect of equivocality are not specific to one particular material or another; they are such as might be exhibited by any of a number of materials. It is a different question altogether from the doctrine of 'truth to material' as that idea seems usually to have been conceived.

Two rather different ideas of truth to material seem to be current. The first has been well summarized in a broadcast by Mr David Thompson. He said that the idea 'in its simplest form means that the sculptor feels obliged to respect his medium to the extent of bringing out in every way he can the stoniness of stone, the metallic quality of metal, the grain and growth and organic properties in wood'.

The second idea is that any given material takes, or can be made to take, certain shapes easily or directly. These unforced shapes are natural to it and are the right shapes to aim at. You must not torture your material.

The two ideas have in common the notion that every material has, as a matter of objective fact, a specific nature, a fixed set of inherent properties, which can be expressed or suppressed when it is used: rather as though it were a child being brought up. They are both essentially concerned with design, and insist that the material shall not be shaped or treated so as to suppress the set of inherent properties which constitute its nature.

The first idea does not tell us how inherent properties are to be expressed but merely that they should be. The second idea, more specific, is that the only, or at any rate the best, way to do it is to make those shapes which come easily.

Let us consider first of all the question of expressing a set of inherent properties.

What are the properties of materials? They are such that they can be defined precisely and measured exactly. The manifest success of the technological revolution proves at least one thing conclusively: that the account which technologists give of the properties of materials is true. Their account is proved true because in every practical application it manifestly works.

They express the properties of materials in figures: ultimate tensile stress, so many pounds per square inch; Young's modulus of elasticity, so many pounds per square inch; weight per cubic foot, such a figure; Brinell hardness, another figure; thermal conductivity, so much; and so on and so on. If therefore the properties of material are to be, or can be, expressed, these are the sort of facts to be expressed.

Some properties of some materials can perhaps be expressed or at least hinted at artistically, but certainly not all properties of all materials. There can be no general principle applicable in all cases. Take, for example, metals. One might well express the ductility of lead; one might even hint at its weight; but not at its low melting-point, not at its electrical resistivity; not at its impenetrability to X-rays, not at its toxic properties. Yet these properties are quite as characteristic of lead as its weight and ductility, and may well matter more. They and a dozen other properties of metals—bronze, aluminium, lead or whatever you please—have as good a right to be expressed as weight and ductility. We are quite arbitrary in ignoring them. Moreover, in some processes, particularly where heat is used, the properties of our materials change under our hand. If, for example, you are forging at the anvil, the hardness and elasticity of your material—either steel or iron—changes with extreme rapidity as it cools. You are thus left with a piece of cold, hard, elastic steel having the visible characteristics of a material with the consistency of wax: for steel at a forging heat has that consistency, and indeed wax is used for carrying out experiments in the technology of rolling steel. Thus the only properties which the smith can express in his finished work are precisely those which the material has lost. The same thing of course is true of roughly modelled clay which has afterwards been fired (plate 7).

Much of the pleasure these things give us comes from the very fact of 'soft' properties being expressed in a hard material to which they are quite foreign. In a similar way, perhaps the most constant and delightful aesthetic phenomenon throughout the history of sculpture has been this very expression in hard stone of the properties of soft materials like flesh, hair and drapery. The stone remains recognizably stone yet the hair is recognizable as hair and the cloth as cloth. If one attempts to discredit the expression of these things then one must maintain that sculpture has only just begun.

Let us turn now to wood. We are to bring out its grain. What does 'bring out' mean? If you develop a negative you bring out the picture. If you warm your brandy glass with your hand you bring out the bouquet. If you take a tube of vermilion and squeeze it you do not bring out the redness of the paint, you merely reveal it: you have not increased or emphasized the redness in any way. Similarly when you cut wood you cannot do anything either to emphasize and express the grain, or to hide it unless by paint. It is there, ready expressed, whether you like it or not. You can, of course, incorporate it into your design; but that has nothing to do with bringing it out and expressing it. The only way one can express the recognizable woodiness of wood is to express the fact that trees are sinuous and branching. Shapes which do not branch cannot express branchingness. Is it seriously to be supposed that wood ought only to be used in sinuous or branching shapes?

The truth is that what we want to do is, not to express the properties of materials, but to express their qualities. The properties of materials are objective and measurable. They are *out there*. The qualities on the other hand are subjective: they are *in here*: in our heads. They are ideas of ours. They are part of that private view of the world which artists each have within them. We each have our own idea of what stoniness is.

This particular idea of truth to material is thus not concerned with objective fact. If we say, 'this is good because it expresses the properties of the material' or 'brings out the woodiness of the wood', we are merely attempting to rationalize our personal preferences. As artists we have every right and reason to express our idea of the qualities of the materials we use, but in doing so we are saying nothing whatever about the materials as they are objectively known.

Let us now consider the idea, that one should do what comes easily. The idea, we said, is that 'Any material takes some shapes directly and easily.

These unforced shapes are natural to it and should be used. You must not torture your material.' The criterion apparently is that the shape shall not have been arrived at in a roundabout way, or else that it shall not have been forced on the material as it were against its will. As for directness, any typical process of the workmanship of certainty used for production in quantity, such as drop-forging, is eminently direct because it does at one stroke what older processes of the workmanship of risk used to do in a much more roundabout and protracted way. Thus if directness be the criterion of truth to material we should eschew the British Museum, for falsehood is rife in it. Almost nothing in it has been made by the direct processes of quantity-production.

The objection will now be made: 'Yes, but ease of making is another matter and it is an objective fact.' That is true. Making is a part of human behaviour and behaviour is observable objectively. *But* the facts also remain that many things which look difficult are easy and things which look easy are difficult: some things which are easy are slow, some things which are very difficult can only be done very quickly. Ease in making is an objective fact, but it is not one that laymen can judge of. If truth to material comes of it alone then he cannot judge of that either.

But suppose we have managed to find out what was easy and what was not. Shall we then reckon the merits of the work proportional to the ease? Is the equation 'easy = good' or 'easy = true'? If so, the primrose path has taken itself a hairpin bend!

Once again, if we say 'this is good because the shape is unforced or direct', we are not referring to any objective fact but merely rationalizing our personal preferences. So that truth to material is merely truth to somebody or other's idea about material.

The people who launched this idea apparently failed to distinguish between material and technique. To be sure there is no general principle of truth to technique but there are some techniques (and only some) which ought not to be substituted for each other unless the form of the thing made is changed also. Wrought iron work is a fair example. Its technique is rather intractable and the range of forms is small which a smith can make without spending an inordinate amount of time. In compensation for this economic restriction, wrought iron work in the hands of a good smith is beautifully free and diversified within its limits. Now a cast-iron imitation of wrought iron work, such as one sometimes sees in London railings, will almost inevitably lose the freedom and diversity, and keep only a highly regulated version of this restricted range of forms; which in themselves have not, to the eyes of our time at least, any extraordinary interest.

Questions of good and bad workmanship do not turn on 'truth to material' or on honesty or deception. Bad workmanship is a matter of making mistakes through hurry, carelessness or ineptitude, which thwart the design: or else of making things look equivocal independently of the design. Of course there are all sorts of deception in workmanship, most of them quite innocent. No one supposes that the lady with pearls as big as birds' eggs round her pretty neck is flaunting the wealth of the Indies, nor would the imitation marble in St Peter's in Rome deceive a child, close to, nor is it expected to. These things are open and rather cheerful bravura, not deception.

Some design goes in for deception about workmanship without quite being fraudulent. Examples of it are the locks on cheap suitcases, which are made to look, more or less, as though of solid brass, but are actually of thin sheet steel, the joints in which can be seen. It is hardly likely that anyone will be taken in for long, but at first glance one might be.

Turning now to the question of equivocality, most kinds of it arise where there are polished surfaces. The idea of polishing has ancient and definite associations which are woven firmly into the fabric of ideas which make our culture. The sight of a polished thing has hitherto been able to activate a complex of associations and attitudes which, consciously or unconsciously, have coloured our feelings about it. That a thing has been polished is as much as to say that special attention has been given to it. Unnecessary work has been done on it, even lovingly. It glitters. It catches the eye. It has an element of excitement about it. It is not dull: nothing can be both dull and polished. A polished surface traditionally implies special significance, something out of the common.

Now in a similar fashion a rough or jagged edge on a piece of finished material unmistakably speaks of abrupt, brutal, contemptuous treatment. To see it is to feel this immediately.

There is thus a total incongruity and a sense of outrage about a piece of material with a highly polished surface and a raw, rough edge. It is ambiguous in the extreme. Moreover, the mere tactile implications of what we see are unpleasant. We imagine the soft ends of our middle fingers sliding gently across the smooth surface and suddenly torn open on the raw edge. This particular deficiency in workmanship is now very widespread. Instances are most usually seen where components are stamped out of sheet metal. The sheered edge is neat and not, perhaps, raw; but it is visibly rough. It is left unmodified and the whole thing is then plated so that it takes a high polish. If special finishing operations were done to clean up the edge before plating the cost would be considerably increased, so they are omitted. The effect is quietly barbarous. It is like that of a highly polished piece of wood with a roughly sawn edge.

Any polished surface which has a mirror-like finish and gives sharp recognizable reflected images, is necessarily likely to produce an equivocal effect because the surface itself is all but invisible and so becomes most difficult for the eyes to bear on. We are thus left in some slight unease because we are uncertain about what we are looking at and where. This unease is attributable to a fact which greatly influences our feelings about the qualities of surfaces. It is that we can have no direct *rapport* with the nature of any material, but have to judge what it is by looking at the surface. We can never see the thing, the material itself, but only the surface, which our vision, unlike X-rays, will not penetrate. Because of this we have habituated ourselves to extracting a surprising amount of information from the look of a surface. From it we judge not only whether a thing will feel rough or smooth, but also whether it will prove to be light or heavy, a good or bad conductor of heat, dry or wet, soft or hard, firm or quaking, coated or 'natural'. We all soon become adepts at this, just as we do at judging mood from the look of a face.

When we find we have seriously misjudged the quality or consistency of a thing at sight we get a shock. It is not only the pain of touching hot iron which upsets us, it is the treachery of the stuff as well; and everyone knows the *frisson* we get from touching something which we thought was hard but find to be soft under a cohesive skin. The expertise we acquire is built up by making comparisons, and we make a judgement about something by considering which it looks like among all the things we have already tested. In this way, for example, we form a generalized notion of what particular 'look' means that a metal thing is tinny and not solid: much as in other parts of the world we shall have formed a notion of what particular 'look' means crocodile and not log—if we know what is good for our health. So it comes about that when we are unable to form any judgement we are somewhat uneasy. This I think explains why we find a mirror polish equivocal.

Silver, when new and highly polished, looks unsatisfactory because of the uncertainty we have been discussing. But after use and wear it comes into its own. Then the network of minute scratches on its surface, although it does not much blur the reflections, provides a visible boundary for the eyes to bear on at the real surface of the material, in front of the reflections and distinct from them. There is, as it were, an unbroken screen of fine scratches and small dents at the surface of the metal, through which we see the reflections.

A similar unbroken screen having a no less satisfactory effect, is provided automatically, without any wear, when wood is brought to a smooth surface and then polished without the use of a lacquer, simply by friction and possibly a little grease, oil, or wax. The small pits and striations of the grain sprinkled over the surface then remain open and establish the desired visible boundary of themselves, unless an exceptionally dense wood like ebony is used. Ebony can be given a particularly revolting high-gloss finish by the use of polishing soap, and because the colour is so dark this gloss completely annihilates the remarkable character of the wood. It is difficult to rub less dense and dark coloured woods to such a high finish and they can usually be spoilt only by lacquer. Very good and thin French polish after some years of use and wear gives a most beautiful quality to the reflections in such woods. The wear is once more necessary to produce a faint, very faint, visible boundary, because the wood will have had its grain filled before polishing.

Wear only improves the quality if the coat remains nearly continuous even though distorted and very slightly abraded. A coat of polish with gaps worn in it looks merely like what it is: a tattered coat fit for a scarecrow.

Unfortunately it is all but impossible to show these distinctions on paper. If the most expert of photographs is put beside the object from which it was taken it becomes instantly apparent that the camera has told less than the whole truth about surface quality. In a reproduction of the same photograph the loss is inevitably greater still. The nuances which have been lost are very subtle, certainly; very difficult to point out, impossible to describe. But so are the nuances of voice and facial expression by which alone a personality is vividly made known, and without which a face is merely an enigmatic mask. Only good workmanship can supply these nuances and without them much of design goes for nothing. It is largely for them and the quality they import into the environment that workmanship is a matter of serious concern. It is almost impossible to demonstrate them adequately in a book. As for the phenomena I have been describing, however, the reader will readily find examples of most of them in almost any building nowadays.

Before we can go farther into the consideration of surface quality, particularly on wood, there are a few facts which must be rehearsed, some of which we have already remarked on:

First. Equivocality in the look of the surface of a thing is apt to make us faintly uneasy, because we expect to be able to make certain reliable judgements from it.

Secondly. When we look at distant things they are stationary, but when we look at things close to they appear to move and change their shape as we ourselves move our head or walk past them. Consequently we arrive at a very correct idea of what they are really like because we see several different versions of them and not merely one; which might well give a false impression because of accidents of light and shade or reflections in polished surfaces. Thus, even when a polished surface appears from a particular viewpoint to be broken because of reflections, we know very well that it is continuous in spite of its appearance.

Thirdly. A recognizable reflection of a thing, technically called a virtual image, from a flat or convex surface having a mirror polish, lies *beyond* that surface. When we bring our pair of eyes to bear on it their lines of sight are converging on something at a distance beyond the surface and not on the surface itself. If we face a mirror we shall find it impossible to look with both eyes at the reflection of a picture on the wall behind us and simultaneously at a fly sitting on the mirror in front of that reflection. They are too far apart for the eyes to bear on both at once, for all that the fly is superimposed on the reflection.

Fourthly. New, highly polished silver produces the same situation as the fly on the mirror. We cannot comfortably look both at the reflections and at the unreflecting parts of the surface adjacent to them because the latter lie nearer to our eye than the reflections.

Fifthly. When the silver gets worn, a continuous screen of minute scratches appears. It is continuous because so many of them show up over the reflected lights as well as over the dark parts of the surface. This screen establishes the position of the surface and makes it clearly apparent. We not only know it must be there, we can see that it is there too. No equivocality remains.

Sixthly. In polished wood without a skin of lacquer—that is to say with an unfilled grain—the pits of the grain form a similar screen.

Now, supposing that the grain of the wood has been filled so that none of the pits remain, the figure of the grain still remains visible although the surface is perfectly smooth. But the lines of the grain and the light and shade in it have the effect of suggesting undulations and ribs in the surface. For example, the rippled figure known as 'fiddle-back' looks like the ripples left on the sand when the tide has fallen, being in fact a flat section cut through fibres rippling in much that fashion. So we are presented with a treble uncertainty. Instead of looking at a continuous smooth surface we are looking at three different things, none of which can be seen for what it is. First there is the surface of the lacquer, which we cannot discern completely because it is too clear. Beyond that we see the wood, which because of the figure of its grain does not look flat although from long experience we realize it must be so. Beyond the wood, in a limbo of their own, are the shadowy but distinct virtual images of things round us and, more distinct, the light reflections of windows or lamps.

The equivocality is most marked when the polished surface is large and flat. In a small object it bothers us much less, because then whenever we look at the surface we also take in its boundaries, and they, being linear, establish the plane of it as effectively as the screen of scratches on silver. Much the same effect is produced by a grid of lines engraved on a highly polished surface, or by inlaid lines. On a cylindrical or other convex surface such as a moulding, the bright reflections are so distorted as to become highlights which have no effect of breaking the surface but, quite to the contrary, accentuate the form and make it easier to discern.

There are two ideally distinct types of reflecting surfaces: specular reflectors, having a high polish, which reflect all the light coming from one direction in a regular way, and, if flat, give mirror-images—virtual images: and diffuse reflectors, such as a sheet of rough white paper, which reflect back the light, but in doing so scatter it in all directions. Now there is obviously a great range of reflectors in between the two. Those at one end of the range give slightly blurred though recognizable mirror-images. Those at the other are surfaces which are almost matt, but have a slight sheen on them. Somewhere about the middle of the range are most lacquered, waxed or oiled wood surfaces and many plastic ones. They are partly specular and at the same time partly diffuse reflectors.

The blurred mirror-images of windows and other light sources in fairly dark-toned surfaces such as these, have an important property. They show up and much exaggerate slight irregularities or roughnesses or undulations in the surface, even though these are quite imperceptible except in these light reflections. At the edges of the light reflections they are shown up most of all.

This means that, because of the light reflections, crevices, pimples, little lumps and rough ridges are suddenly revealed in surfaces which we thought were smooth, and which elsewhere look smooth and indeed are smooth by ordinary reckoning. In other cases the surfaces we took for flat turn out to undulate or have wide shallow dents and dimples. The surface qualities which we judge unpleasant, without as a rule considering why, are often those which are shown by the edges of the light reflections to be still in need of cleaning off or smoothing down. Our aversion turns as usual on the feeling that an intention has not been carried out; that something has not been done which ought to have been; that the surface has been 'polished' to make it look smooth and cared for, but then we find we have been cheated and it is not smooth at all. The thing is ambiguous, equivocal. It blows hot and cold with the same breath.

It may be also that the tactile implications of what we see trouble us. A thing may look smooth in one light and rough in another, but it can never feel both at once.

Just as no surface has ever been flat and no angle square, so no surface can be smooth or anything approaching it. But things can look smooth, and it is that which concerns us. What looks to us a smooth matt surface is something more like a ploughed field; but the separate furrows and clods which compose it are so small that the eye is unable to resolve them—to see them as separate. The subjective quality of a surface, its beauty in the eye of the beholder, alters strangely when it is magnified two or three times. As we have seen when considering what I have called Diversity, that which looks good at one scale may well look bad at another and will certainly look different. When we see a smooth surface we see an agglomeration of separate objects which are just beyond the threshold of resolution of the eye. Smoothness in fact is a sensation and an appearance; but these do not necessarily correspond to one objective definable condition of the surface. Surfaces are more smooth or less; never absolutely smooth. The workmanship of surface finishing is an affair of producing an illusion of smoothness in any case, and the defects of surface revealed at the edges of the light reflections are themselves largely illusory or at least greatly exaggerated. But the fact that they are illusory does not mean that they are trivial. The whole world of appearances is a world of illusion on various levels, but what those illusions mean to us consciously and subconsciously is a prime fact of experience and the quality of our life is immeasurably affected by it.

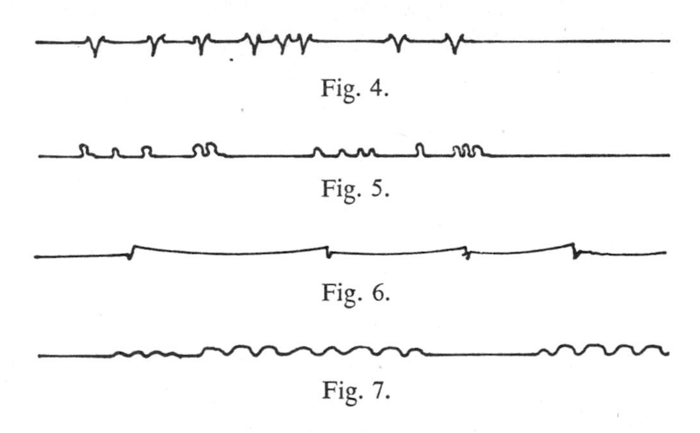

Fig. 4.

Fig. 5.

Fig. 6.

Fig. 7.

In the light of this it may be useful to consider in detail some surface qualities of wood. Fig. 4 shows a section through a lacquered surface drawn not as it actually is but so as to show the sort of roughnesses that the light reflections suggest are there. The skin of lacquer seems to have cracked along the lines of the grain or else built up at the edges of the minute crevices of it, so that each of them has raised edges. Fig. 5 shows a surface which appears to have (and probably has) minute whiskers on it, each of which has gathered a little clot of lacquer. Fig. 6 shows the effect of defects in marquetry which add nothing to the quality of old furniture. Fig. 7 indicates the effect caused by a defect in lacquer known as 'orange-peel'; an irregular pattern of crinkles and dimples such as its name implies. As a result of these, the edges of the light reflections are broken up into an irregular spatter of dots and patches which are the highlights exaggerating the little lumps of the surface. Fig. 8 shows a surface of some wood, such as Rio rosewood, having open pits in the grain, which has been brought to a good smooth surface.

Just as quite a low polish will show up and exaggerate minute defects, so also will it emphasize the smoothness of a surface which really is smooth however we touch or look at it. In a close-grained wood such as beech, which has no open pits to provide the screen-effect if it is polished, the figure of the grain will itself replace them, so long as the polish is not too high and the figure remains visible through the dimmed reflections of the lights as well as in the unreflecting parts: that is to say, strictly, the parts where there are reflections of dark objects which can hardly be made out.

A worn or distressed surface of good workmanship has the character of fig. 9: The essential overall flatness and smoothness are unbroken. The dents seem to be fairly steep but not sharp.

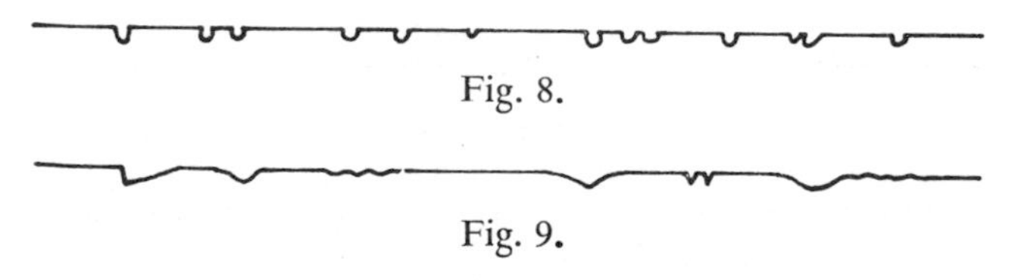

Fig. 8.

Fig. 9.

There is nothing ambiguous or unpleasant about a surface where the same pattern of unevenness can be seen both at the edges of the light reflections and elsewhere and where the unevennesses are themselves evenly polished. A great variety of such surfaces is found in seashells where the thinnest possible coat of very shiny transparent glaze is laid over a hummocky or reticulated or furrowed or striated or minutely sculptured ground, which is often patterned in colour also. In some sea shells, as in the shells of ordinary hens' eggs, only the tops of the hummocks are polished, while the pits and valleys in between them are dull. The same thing can be seen in worn hand-smooth wood when the surface has scoured out, leaving the hard summer rings of the grain upstanding. In all such cases the mysterious attraction of a polish is combined with the faint sense of assurance given by a visible surface defined by the screen of minute highlights on the hummocks, and so there is no uncertainty or ambiguity. A rather unevenly plastered wall with high-gloss paint on it is yet another example of an uneven polished surface with a pleasant character. It leaves us with no sense of careless finishing or inadequate smoothing. The unevenness of the wall is consistent and regular: an even unevenness. We no more question it than we should the uneven veining in marble.

We can learn to measure and discriminate by eye with astonishing accuracy, and in the workmanship of risk the workman relies very much on this ability. We should never be done if we had to check the size and shape of every minutest thing by instruments. Everyone is good at detecting departures from straightness and flatness, and when we make judgements based on the appearance of surfaces we manage to extract a good deal of information from certain particular kinds of near-flatness or undulation revealed by the light reflections on polished surfaces, because we habitually associate these with distinctive properties of things. We know the surface which has a tinny look, and the one which looks like a coat of paint imperfectly attached to its ground, curling at the edges and blistering in the middle: and the one which looks like a congealed skin over the surface of mud or porridge, faintly crumpled: and the one which looks leathery and another which looks papery.

They *look* tinny or leathery or whatever it may be. Two kinds of equivocality can result: one of the sort we have already discussed. This we see when at first sight the lid of a tin or the side of a ship seems even and flat; but when the light glances across it we see that the tin-lid is buckled and every frame of the ship grins through the plates along her side. We find we have been, however innocently, deceived. The second kind of equivocality is perhaps rather a matter of incongruity and there is no element of deception in it. It is seen in certain solid thermosetting plastic mouldings such as electric power-plugs. The profile of these, and in some cases various ribs formed on the surface, are most precise and rigid looking; moreover some of the arrises are very sharp and, on a minute scale, raw. But the flat surface in between is not so precise and has something of a soft leathery quality, showing undulations and buckles when the light strikes across it. So we have a thing whose surface has a smooth soft vague quality, but the edges, details and profile of it have a sharp precise rigid look about them.

All that I have said here relates to kinds of equivocality produced by faulty workmanship which are immediately related to objective phenomena and which can at least be demonstrated to all observers alike. But the subjective importance of surface qualities is another matter. They are peculiarly the workman's preoccupation. Only he can control them. Aesthetically they matter not less than colour. If all the colours of noon had the same surface quality they would horrify us; for colours take half their life and interest from the quality of the surface to which they are applied or in which they inhere. Again, consider the difference between the surface of an eggshell and sharkskin, a rose petal and velvet, ivory and soap, a peach and a baby's skin. We have few enough names for colours but for surface qualities all but none. Yet the variety of our experience of surface quality must be every bit as wide as that of colour.

The extreme paucity of names for surface qualities has quite probably had the effect of preventing any general understanding that they exist as a complete domain of aesthetic experience, a third estate in its own right, standing independently of form and colour. If that is not so, what is it that we see in black-and-white photographs? Nothing can ever be seen anywhere except surface; we can never see more of material things than that unless they are transparent or translucent. If a good black-and-white photograph did not exhibit surface quality, similarity of tone in it would imply similarity of material.

Surface quality in man-made things comes of workmanship. The third estate belongs to workmanship.

10

Critique of 'On the Nature of Gothic'

Most ideas of workmanship now current in the West, such as they are, have been coloured by the doctrines of the Arts and Crafts movement. Those doctrines have never been distinctly formulated, so far as I know, perhaps because they cannot be. The influence they have had and still exercise makes it necessary to examine them.

The ideas which launched the movement seem to have been Ruskin's. He supposed that his aesthetic preferences and his social aims supported each other, and that more of the workmanship he liked to see would mean more happiness for the workmen who did it. These ideas must have been written down by 1850, and appeared in the famous chapter on 'The Nature of Gothic' in *Stones of Venice*. The chapter contains about 30,000 words and amounts rather to a book.

Morris in his lecture on 'The Lesser Arts', in December 1877, says

As for the last use of these arts, the giving us pleasure in our work, I scarcely know how to speak strongly enough of it; and yet if I did not know the value of repeating a truth again and again, I should have to excuse myself to you for saying any more about this, when I remember how a great man now living has spoken of it: I mean my friend Professor John Ruskin: if you read the chapter in the second volume of his *Stones of Venice* entitled, 'On the Nature of Gothic, and the Office of the Workman therein', you will read at once the truest and most eloquent words that can possibly be said on the subject. What I have to say upon it can scarcely be more than an echo of his words...

Eventually the chapter was published as a separate book. In 1892 Morris said in a preface which he wrote to it:

...To my mind, and I believe to some others, it is one of the most important things written by the author, and in future days will be considered as one of the very few necessary and inevitable utterances of the century. To some of us when we first read it, now many years ago, it seemed to point out a new road on which the world should travel. And...we can still see no other way out of the folly and degradation of civilisation. For the lesson which Ruskin here teaches us is that art is the expression of man's pleasure in labour...

Let us consider a few of Ruskin's words from this chapter which perhaps will convey some of the substance of his thought even in isolation:

§ *xii.* Men were not intended to work with the accuracy of tools, to be precise and perfect in all their actions. If you will have that precision out of them you must unhumanise them.

If you will make a man of the working creature you cannot make a tool. Let him but begin to imagine, to think, to try to do anything worth doing, and...out come all his roughness, all his dullness, all his incapacity...failure after failure...but out comes the whole majesty of him also.

§ *xiii.* And now, reader, look round this English room of yours, about which you have been proud so often, because the work of it was so good and strong, and the ornaments of it so finished. Examine again all those accurate mouldings, and perfect polishings, and unerring adjustments of the seasoned wood and tempered steel. Many a time you have exulted over them, and thought how great England was, because her slightest work was done so thoroughly. Alas! if read rightly, these perfectnesses are a sign of slavery in our England...

§ *xiv.* Gaze upon the old cathedral front...those ugly goblins...and stern statues anatomiless and rigid...are signs of the life and liberty of every workman who struck the stone.

§ *xv.* The degradation of the operative into a machine...It is not that men are ill fed, but that they have no pleasure in the work by which they make their bread and therefore look to wealth as the only means of pleasure.

§ *xix.* For observe, I have only dwelt upon the rudeness of Gothic, or any other kind of imperfectness, as admirable, where it was impossible to get design or thought without it. If you are to have the thought of a rough and untaught man, you must have it in a rough and untaught way; but from an educated man...take the graceful expression and be thankful.

§ *xix.* Above all demand no refinement of execution where there is no thought, for that is slaves' work, unredeemed. Rather choose rough work than smooth work so only that the practical purpose be answered, and never imagine there is reason to be proud of anything that may be accomplished by patience and sandpaper.

§ *xxi.* On a large scale, and in work determinable by line and rule, it is indeed both possible and necessary that the thoughts of one man should be carried out by the labour of others... But on a smaller scale, and if a design cannot be mathematically defined, one man's thoughts can never be expressed by another: and the difference between the spirit of touch of the man who is inventing, and of the man who is obeying directions, is often all the difference between a great and a common work of art.

§ *xxi.* It would be well if all of us were good handicraftsmen in some kind, and the dishonour of manual labour done away with altogether.

§ *xxii.* But, accurately speaking, no good work whatever can be perfect, and *the demand for perfection is always a misunderstanding of the ends of art.* [Ruskin's italics.]

§ *xxv.* Nothing that lives is or can be rigidly perfect; part of it is decaying, part nascent.

In all things that live there are certain irregularities and deficiencies which are not only signs of life, but sources of beauty.

...to banish imperfection is to destroy expression, to check exertion, to paralyse vitality.

§ *xxvi.* I have already enforced the allowing independent operation to the inferior workman, simply as a duty to *him*, and as ennobling the architecture by rendering it more Christian...

§ *xxix.* ...change or variety is as much a necessity to the human heart and brain in buildings as in books... there is no merit, though there is some occasional use, in monotony.

The quotation at *xv* 'It is not that men are ill fed...' and the one at *xxi* 'the difference between the spirit of touch of the man who is inventing...' seem admirable but some of the others not so.

Ruskin was a man of great insight and a great writer. A passage at *viii* in this chapter is a fair example of his artistry and imaginative power: it begins '...but we do not enough conceive for ourselves that variegated mosaic of the world's surface which a bird sees in its migration, that difference between the district of the gentian and of the olive which the stork and the swallow see far off, as they lean upon the sirocco wind'. His deep belief in the serious importance of art and the sensitivity of his perception must have opened many eyes as they once did mine. Ideas born in his mind have had an immense influence and those of them which bore on workmanship are active still. Some of these ideas were by no means his best.

It is necessary to insist on his stature because his shortcomings were fairly plentiful and in the present context I am more concerned with them than with his great merits.

He preferred rhetoric to the exact analysis of ideas, and much preferred it to the definition of his terms. He did not try to use words exactly. The reference of words like 'rough' and 'perfect' is hardly ever certain. He could not write 'seasoned wood' without adding 'tempered steel' so as to balance it. But there is no analogy between tempering and seasoning, and there is no tempered steel in 'that English room of yours' except in the dinner-knives and the clock-spring. More particularly, and most importantly in the chapter we are considering, he never managed for long to dissociate the idea of workmanship from that of carving ornament.

The chapter on 'The Nature of Gothic' continues the theme of an earlier one on 'Treatment of Ornament', in which he says (*xiii*)

> This is the glory of gothic architecture, that every jot and tittle, every point and niche of it, affords room, fuel, and focus for individual fire. But you cease to acknowledge this and you refuse to accept the help of the lesser mind, if you require the work to be all executed in the grand manner. Your business is to think out all of it nobly, to dictate the expression of it as far as your dictation can assist the less elevated intelligence: then to leave this, aided and taught as far as may be, to its own simple act and effort; and to rejoice in its simplicity if not in its power, and in its vitality if not in its science.

The statement about jots and tittles is untrue even on the evidence of the illustrations he himself drew for his book—unless by architecture he means ornament. You cannot get individual fire into plain walling or the cylindrical shaft of a column. Moreover, from time to time he half acknowledged its untruth. Yet he continually returns to this obsession and writes accordingly. He writes as though building were ornament and, by extension, as though workmanship were almost synonymous with ornament.

The deficiencies of the Arts and Crafts movement can only be understood if it is realized that it did not originate in ideas about workmanship at all. Indeed it never developed anything approaching a rational theory of workmanship, but merely a collection of prejudices which are still preventing useful thought to this day.

Much of what Ruskin writes is ambiguous because it is impossible to be sure what he is referring to. When he cites examples he always manages to leave room for doubt about his meaning. So far as one can judge, the essence of the ideas he wanted to express was that:

> To make men do tedious repetitive tasks is unchristian.
>
> High regulation always involves such tasks and must therefore be eliminated.
>
> If the workman is allowed to design he will do rough work and so will eliminate it.

Above all, the workman's naïve designs will be admirable. What Ruskin was inveighing against was not hard labour, but patient work. He did not realize, or so it seems, perhaps because he had never had to work for a living, that a fair proportion of patient tedious work is necessary if one is to take pleasure in any kind of livelihood, whether it be designing or making, for no one can continuously create and no one ever has. He did not realize there is great pleasure in doing highly regulated workmanship.

He was making propaganda for a certain strain of naïve ornament and for free workmanship (as I have termed it). He persuaded himself and Morris that by so doing he was offering a cure for the miseries of industrialization.

He did, however, realize that the life of his times depended on highly regulated workmanship for its continuance, and when he formulated his precepts he was obliged to make exceptions which relegated his teaching to the periphery of industry: though neither he nor his followers seem to have been ready to admit this.

So far, again, as one can understand him, a fair summary of what he actually wrote in this chapter is as follows:

1. Men can only take pleasure in their work if they are allowed to invent, to exercise thought: that is to say, to design as well as to make.

This is, of course, a dogma and he gives no evidence in support of it. Perhaps what he would say is 'Men *ought* only to take pleasure...'

He seems to have meant them to design as they worked, not before they started.

2. It is a Christian duty to allow the men to do so.
3. The workmen through no fault of their own are untaught, unsophisticated.

4. Therefore, because of their 'incapacity' their designs will be rough and imperfect.

By imperfect he probably meant naïve, incorrect in rendering anatomy and similar matters.

5. Their imperfect designs will be admirable because of their imperfection. 'Of human work, none but what is bad can be perfect in its own bad way.'
6. They will, if required to 'think', i.e. to design, *necessarily be incapable* of giving an exact or perfect finish to their work; and such a finish is a sign of slavery. The workman is degraded by being required to give it. (Exact or perfect finish means, in my terminology, high regulation.)

This last tenet, that he who designs is thereby made incapable of 'perfect' work, was apparently a matter of absolute dogma. He gives no reason why it should be so, or example to demonstrate that it must be so; but every carver knows that it is not so, and in a trade like joinery or cabinet-making countless examples demonstrate that it never has been so. Yet, (*xx*) when praising rough workmanship in Venetian glass, he states it explicitly: 'If the workman is thinking about his edges, he cannot be thinking of his design; if of his design he cannot think of his edges. Choose whether you will pay for the lovely form or the perfect finish, and choose at the same moment whether you will make the worker a man or a grindstone.' Because blown glass has a quality impossible to cut glass (which he disliked) it does not follow that accurate workmanship is impossible to a workman who designs. This, moreover, is Ruskin's only doctrine, in this chapter at least, which can confidently be said to refer to workmanship rather than the design of ornament.

In 'that English room of yours' how did he know that the workman had *not* designed the joinery, with its 'unerring adjustments of the seasoned wood'? Did he really suppose that it could ever have been designed by the workman while he was at work? And since it is flatly impossible to do that, as any workman knows, and since an accurate drawing has to be made first on a rod; how is the workman more enslaved by working 'perfectly' from another man's drawing than he is by working from his own?

The truth of it is, that Ruskin, as usual, is asserting that if he does not like something, it must therefore be thoroughly evil. 'Salvator Rosa and Caravaggio...perceive and imitate evil only' (*liv*). Because he liked rough workmanship, high regulation was therefore evil, and there was no need to stop and think why. Because he did not like an early Victorian room, everything about it, workmanship included, must be evil.

He sees and eloquently condemns the evils which resulted from the industrial practices of his day. He says (*xvi*) that 'in our manufacturing cities we manufacture everything except men': that 'to brighten, to strengthen, to refine, or to form a single living spirit, never enters into our estimate of advantages'. He then says that 'this evil can only be met by a right understanding on the part of all classes of what kinds of labour are good for men, raising them, and making them happy; by a determined sacrifice of such convenience, or beauty, or cheapness as is to be got only by the degradation of the workman; and by an equally determined demand for the products and results of healthy and ennobling labour'.

And how, he asks, are these products to be recognized and this demand to be regulated?

Easily, by the observance of three broad and simple rules:

(1) Never encourage the manufacture of any article not absolutely necessary, in the production of which *Invention* has no share.

(2) Never demand an exact finish for its own sake, but only for some practical or noble end.

(3) Never encourage imitation or copying of any kind except for the sake of preserving record of great works.

So far Ruskin. At once we see that in (1), if an article is absolutely necessary, then we need not discourage its manufacture even though invention has no share: and in (2) we may rightly demand an exact finish if for some practical end. *But* the central idea of his teaching and of the Arts and Crafts movement was that it is wrong to deny the workman the opportunity of inventing, and that if he is required to produce an exact finish under someone else's design it is slavery. Why, then, does wrong become right and slavery freedom as soon as a necessary or practical end are in view?

Presumably Ruskin has realized that if he does not make this exception he is condemning most of the rapidly expanding population to starve. In his day the greater part of manufacture was directed to the strictly practical ends of making such things as bricks, tiles, slates, boards, castings, rails, forgings, ships, machinery, vehicles, warehouses, docks, roads, tools, factories, mills, agricultural implements. Where in all that is the workman to do his inventing unless the clock is put back a hundred years? So, much the greater part, and the essential part, of industry must be made an exception to Ruskin's rule. He has no remedy to offer for the manifest evils of any industry which caters for the practical necessities of life, only for what is inessential!

This, I imagine, has for long been an accepted criticism of Ruskin's ideas and of the Arts and Crafts movement; and it is a damning one. But what, I think, is not realized is that the workmanship he is condemning so strongly because it degrades the workman, such as that which he describes in 'that English room of yours', is the workmanship of risk. It is precisely what today would be described as the finest craftsmanship. What he is against is *not* the workmanship of certainty, or quantity-production by machine-tools, which indeed he scarcely alludes to, but high regulation.

Now in the middle and latter part of the nineteenth century high regulation in the workmanship of risk was brought to a pitch which presumably we shall never see again on any scale. The harm which this chapter on the Nature of Gothic has done is great. In it Ruskin did injustice to Victorian workmanship and to the men who produced it, whom he called slaves; and he influenced William Morris to cause yet more harm. Between them they diverted the attention of educated people from what was good in the workmanship of their own time, encouraged them to despise it, and so hastened its eventual decline. Morris, in his article on 'the Revival of Handicraft' already quoted here on page *12*, writes:

With those who do understand what beauty means I need not argue it, as they are but too familiar with the fact that the produce of all modern industrialism is ugly, and that whenever anything which is old disappears its place is taken by something inferior to it in beauty; and that out in the very fields and open country. The art of making beautifully all kinds of ordinary things, carts, gates, fences, boats, bowls, and so forth, let alone houses and public buildings, unconsciously and without effort, has gone; when anything has to be renewed among those simple things the only question asked is how little it can be done for, so as to tide us over our responsibility and shift its mending onto the next generation.

He might perhaps have produced at least a show of justification for these remarks in their application to bowls, buildings, and even fences. But as to country carts, gates and boats, he is implying that they were in 1888 the products of modern industrialism: which was untrue then, and was still untrue fifty years or so afterwards. He is also saying that those of 1888 were uglier than those which they replaced, whereas in very many cases the same traditional designs and makes continued then and thereafter: while in the great majority of others the changes were evolutionary and often, we may think, for the better (but here it must be remembered that much of what we should call good design, he would have called 'utilitarian ugliness'). Worst of all, and unforgivable, he implies unmistakably that these things were shoddy ('...how little it can be done for...tide us over our responsibility and shift its mending...'). As for this, many ordinary things of that age have survived to ours, and by them Morris's untruth stands condemned.

The things which the Arts and Crafts movement produced may be thought to have justified it. Some of the architects and designers influenced by it did work of the greatest distinction, not all of which is yet appreciated; and Morris himself was, of course, an unequalled designer of patterns. But the pretensions of the movement are difficult to forgive, for people who believed in it pretended that its doctrine of workmanship was the only true one. That inchoate doctrine will not stand fire. The movement neither formulated it precisely nor criticized and corrected it in its original form—Ruskin's. Because of this the movement left behind it confusion of thought about workmanship: or, in its terms, craftsmanship. There is to this day no agreement about what constitutes craftsmanship; nor is there any about what is not craftsmanship, and that is perhaps still more significant. One has known craftsmen whose ideas have been coloured by the Arts and Crafts movement, to imply that not-craftsmanship is:

Imprecise workmanship (i.e. rough or free workmanship).
and/or Precise workmanship (i.e. regulated workmanship).
and/or Unskilful work: skill not being defined.
and/or Working to another man's design: or (I think) a traditional design, unless of a musical instrument.
and/or Using machine tools (if they are power driven).
and/or Producing a series of more than perhaps six things of the same design.
and/or Not making the whole job from start to finish oneself.

It should be particularly noticed that, with rare exceptions, you cannot tell, simply by looking at the work, whether the last four criteria apply or not: whether it is the work of a 'craftsman' or not. Consequently these last four ideas have fostered the extraordinary notion that craftsmanship should not be judged by its results like all other workmanship, and that the craftsman may properly take the standpoint 'I am holier than thou'. This attitude is presumably involved to some extent in the conception of the 'Fine Crafts', which, however, have not been defined.

One can only reply to this that a workman who will not be judged by his work is contemptible, and that there is no possible criterion of workmanship except the work. If that too is a dogmatic assertion, at least it has the backing of ancient tradition! 'The tree is known by its fruit.'

Far too much work propagated by the Arts and Crafts movement was either made by first-rate workmen trained outside the movement according to the traditions of the trades—Ruskins's 'slaves', to whom due credit has not been given: or else by inferior workmen trained inside it, who were prepared to invoke the spirit, or the way in which they worked, as an excuse for their ineptitude. It is told of a potter that when reproached about a teapot he said '. . . but of course it leaks. It's hand made.' It is fair to add that there still are in this country some admirable workmen who were brought up in the atmosphere of the movement, who are by no means of that kind. But most of these would agree, I think, that a workman stands to be judged solely by his work.

Before leaving this subject two questions must be asked: in what circumstances do men actually take pleasure in their work? and, what, in spite of the errors he propagated, did Ruskin contribute positively to the theory of workmanship?

The first question is a matter largely of fact and there will, one day, be evidence enough to answer it fully. In the meanwhile one can make a fairly confident guess. In the first place, obviously, pleasure in work depends on not being over-driven or over-driving oneself because of poverty. Given a reasonable wage, a reasonable master and reasonable hours, some people like work which is mindless, repetitive and monotonous, and entirely devoid of risk; but whether that liking can ever amount to pleasure is debatable. Others can only take pleasure in work which, because it involves dexterity and conscious judgement, does involve risk and is not mindless. Such work may, however, be repetitive, and if it is so repetitive that judgement becomes entirely unconscious, then it is debatable whether there can be any positive pleasure in it. Certainly, however, there can be a certain pleasure in finding that one's judgement is being exercised only half consciously and in letting the process continue. I suppose that in many trades where the workmanship of risk prevails, any competent workman does much of his work like that. One can, for instance, do a great deal of sawing and chopping without quite knowing how one has arrived at the result correctly. The hands appear to do it on their own, without referring to the head.

Trades differ enormously, and the degree of risk and amount of dexterity and judgement they require differ correspondingly. There is a strong sporting element in some workmanship. There are times when one can irretrievably spoil in a matter of seconds the work of a whole day. The element of risk is no figure of speech. In such a trade as the blacksmith's the critical moments are also dramatic, as anyone must agree who has watched a fire-weld being made. As the iron comes to the heat the fire roars, the fan hums and the smith stands silent. Suddenly, like an irrupting comet the iron is swept white-hot out of the fire on to the anvil, with scale spattering from it in a blinding shower, and three decisive hammer blows have made the weld. Or not! The timing and control of those movements have decided whether the weld is sound. Many lives on many occasions must have depended on their timing in forging the ironwork for sailing ships. A 'cold shut' or a weld with dirt in it could remain undetected for years and then perhaps bring down a mast, or, if in an anchor, put a ship ashore.

But there are admittedly many things in the workmanship of risk which are to be achieved only, as Ruskin said, by patience and sandpaper; and the satisfaction one gets from them is mainly in seeing the finished job: but in contradiction of what he said, one may on occasion be proud of it.

Is it true that to invent and design the work adds to the pleasure of making it? In work like carving, where it is possible to improvise as one goes along, the opportunity to invent according to one's fancy does indeed give pleasure, if pleasure is the right word for forgetting oneself, and all the world, and time; but there can be, and can have been, few such kinds of work. As for other kinds of invention, I am not so sure. To have designed something and to have made it and seen that it was worth the trouble, certainly gives pleasure. So it may also to be making what one has designed, and to be seeing it take shape; but there is always a certain anxiety in that, and working to one's own design does not necessarily give more pleasure than working to a traditional design, as one does when making tools, or to another man's design so long as it is a good one. The act of interpreting a design and seeing how it turns out can be a great pleasure. It is certain that workmanship can provide many kinds of pleasure that are not at all diminished by not having made the design oneself.

I do not propose to speculate why Ruskin and Morris did not try to find out who, in their day, was actually taking pleasure in what work: or, for that matter, how much designing a medieval workman did do in his short life, in proportion to how much repetitive work. It remains therefore to say what positive contribution to the theory of workmanship Ruskin seems to have made.

I think he propagated three important ideas. He saw, before Japanese aesthetics were known in the West, that free and rough workmanship have aesthetic qualities which are unique. He also saw that in manufacture and building there is a domain of aesthetic qualities which are beyond the control of design, and insisted that architects with drawing boards could never have made Venice what it was. Thirdly he described and understood the quality in things which I have termed diversity (chapter 7) and understood its importance in the design of ornament, though not in workmanship. The intrinsic importance of these ideas is not diminished by the fact that so much rubbish has derived from illegitimate extensions of them.

11

The aesthetic importance of workmanship, and its future

In the foregoing chapters it has been suggested that the importance of good workmanship in its aesthetic aspect rests on three things:

(1) Highly regulated workmanship shows us a thing done in style: an evident intention achieved with evident success. It is anti-sordid, anti-squalid and contributes to our morale.

To do a thing in style is to set oneself standards of behaviour in the belief that the manner of doing anything has a certain aesthetic importance of its own independent of the importance of what is done. This belief is the basis of ordinary decent behaviour according to the customs of any society. It is the principle on which one keeps one's house and one's person clean and neat, and so on. Regulation which, in general, the workmanship of risk can only achieve by taking a good deal of avoidable trouble, used undoubtedly to be a part of this idea of behaviour.

With the workmanship of certainty it is becoming easier to achieve high regulation and less determination is needed to do it; but still the quality of the result is clear evidence of competence and assurance, and it is an ingredient of civilization to be continually faced with that evidence, even if it is taken for granted and goes unremarked.

(2) Free workmanship shows that, while design is a matter of imposing order on things, the intended results of design can often be achieved perfectly well without the workman being denied spontaneity and unstudied improvisation. This perhaps has special importance because our natural environment, and all naturally formed or grown things, show a similar spontaneity and individuality on a basis of order and uniformity. This characteristic aspect of nature, order permeated by individuality, was the aesthetic broth in which the human sensibility grew. Whereas in the early days of civilization highly regulated workmanship seemed admirable

because it was rare, difficult, and exceptional, that situation is now completely reversed, and we might well try to make ourselves an environment which had more concord with our natural one.

(3) Good workmanship, whether free or regulated, produces and exploits the quality I have called diversity, and by means of it makes an extension of aesthetic experience beyond the domain controlled by design, down to the smallest scale of formal elements which the eye can distinguish at the shortest range. Diversity on the small scale is particularly delightful in regulated workmanship because there it maintains a kind of pleasantly disrespectful opposition to the regulation and precision of the piece seen in the large: as when, for instance, the wild figure of the wood sets off the precision of the cabinet-work. Diversity imports into our man-made environment something which is akin to the natural environment we have abandoned; and something which begins to tell, moreover, at those short distances at which we most often see the things we use.

What changes can one foresee? Is there for instance any reason for the productive part of the workmanship of risk to continue doing highly regulated work? Why should it, when the workmanship of certainty is capable of higher regulation than ever was seen? Why, in particular, should it, considering that high regulation by the workmanship of risk is usually very expensive even where the best and most ingenious use is made of machine tools? Imagination boggles at the thought of what it might cost to build any standard family car from scratch by the workmanship of risk. How many weeks would it take to make the carburettor, for instance, or one of the head-lamps?

It should continue simply because the workmanship of risk in its highly regulated forms can produce a range of specific aesthetic qualities which the workmanship of certainty, always ruled by price, will never achieve. The British Museum, or any other like it, gives convincing evidence of that. And one need not copy the past in order to perpetuate those qualities. People still use oil-paint, but they do not imitate Titian.

There is of course no danger that high regulation will die out in the preparatory branch of the workmanship of risk. Beyond that, the prevalence and immense capability of the workmanship of certainty will ensure that highly regulated workmanship continues and increases. Indeed there is already too much of it or, rather, there is too little diversity in it. The contemporary appetite for junk and antiques may partly be a sign of an unsatisfied hunger for diversity and spontaneity in things of everyday use. I do not think it can be quite explained either by the romantic associations of mere age or by an aversion from the ephemerality of contemporary designs. There is still comparatively so much diversity about that it is difficult to estimate how an environment quite devoid of it would strike us. The quality in design which is called 'clinical' is more or less the quality of no-diversity. A little of it, for a change, is pleasant, but a world all clinical might be fairly oppressive, and such a world of design and workmanship without diversity is decidedly a possible one, now.

Four things are going wrong:

1. The workmanship of certainty has not yet found out, except in certain restricted fields, how to produce diversity and exploit it.
2. Where highly regulated components are fitted and assembled by the workmanship of risk, in industries which are only in part 'industrialized', such as joinery for buildings, some of the workmanship is extraordinarily bad.
3. Some kinds of workmanship, such as the best cabinet-making, which use the workmanship of risk to produce very high regulation and the most subtle manipulations of diversity, are dying out because of the cost of what they do. But what they do has unique aesthetic qualities.
4. Free workmanship also is dying out, for the same reasons, and it also has unique aesthetic qualities for which there can be no substitute.

It is, I submit, quite easy to see what might be done about the last three of these things but not about the first, which is undoubtedly the most important. The workmanship of certainty can do nearly everything well except produce diversity. Its only real success in that way at present is in weaving and in making things of glass or translucent or semi-translucent plastics such as nylon or polythene which show delightful diversification because of their modulation of the transmitted light and the interplay between it and the light reflected from their surfaces. Diversity in shapes and surfaces could also, no doubt, be achieved fairly crudely by numerically controlled machine tools, and perhaps something more can be hoped for there in course of time.

Much of the diversity in highly regulated work produced by the workmanship of risk used to be achieved through the manner in which it made use of the inherent qualities of natural materials. It is very probable that, if diversity were appreciated as much as economy, synthetic or processed materials would be made with an equally rich inherent diversification.

If industrial designers and architects understood the theory and aesthetics of workmanship better, and realized the importance of it, they would surely make better use of the opportunities offered by the techniques which are now available to them. One could almost believe that some industrial designers only know of two surface qualities, shiny and 'textured'; and that to them texture means something which has to be distinguishable in all its parts three feet away! They ought to reflect that so far as the appearance of their work goes its surface qualities are not less important than its shape, for the only part of it which will ever be visible is the surface.

The want of diversity is not so much to be blamed on the technologists as on the designers, who do not think enough about it, or do not think enough of it. Perhaps I think too much of it, but it is high time somebody spoke up for it. Art is not so easy that we can afford to ignore any and every formal quality which will not go on to a drawing board. Yet, the fact remains, I can offer no better suggestion than that, if people came to love diversity, they would find out ways of producing it.

The answer to the second problem, of bad workmanship in assembly and finishing off, is much easier to see. The first thing to be grasped is that the situation now is fundamentally different from what it was in the old days of good rough workmanship. The second thing is that the force of the long traditions of the workmanship of risk is now very weak in many trades. With some honourable but rather few exceptions, it no longer concerns a joiner's self-respect and standing in the eyes of his trade, that his work shall be done properly according to those traditions, and moreover he will be paid as well as before even if it is done badly.

This situation is regrettable, but it does not necessarily mean that the joiner is a bad man. It merely means that his education in his trade has been bad (for a trade learnt according to the traditions was an education, though a circumscribed one. It taught the principles on which one should act in certain circumstances and the difference between good and bad actions). The existing situation arises from the fact that the building trade is in transition in this country from the workmanship of risk to that of certainty, to the assembly of prefabricated components so made that neither care, knowledge nor dexterity are required for their assembly; and such trades as the joiner's are in decline. There are now too few good joiners.

It is futile to hope that the process of decline can be reversed on a sufficient scale to match the size of the industry, and the action to be taken is unmistakable. We must stop designing joinery and other details of cheap buildings as though for such work we could command fully educated joiners whenever we wanted them. It is, for example, silly to design architraves which have to be mitred round door openings. Of all joints a mitre is sure to be badly done or to go wrong in cheap work. It is necessary for the architect to understand very clearly the limitations of the workmanship which the price of the building will allow, to understand that nothing can be left to the discretion of men without education in the trade, and to design within those limitations instead of asking for highly regulated traditional joinery like mitred architraves.

As for the third and fourth problems it is again not difficult to see a line of action, but it may not be easy to arouse interest and inform opinion so that the action gets taken. It will be a great loss to the world if at least a little highly regulated work does not continue to be done by the workmanship of risk in making furniture, textiles, pottery, hand-tools, clothes, glass, jewellery, musical instruments and several other things. It will equally be a loss if free workmanship does not continue. Most of such work will fall within the province of what are now called 'the Crafts'. What is now required is a more realistic conception of them.

The workmanship of risk can be applied to two quite different purposes, one preparatory, the other productive. Preparatory workmanship makes, not the products of manufacture, but the plant, tools, jigs and other apparatus which make the workmanship of certainty possible. Productive workmanship actually turns out products for sale.

The preparatory branch of the workmanship of risk is, of course, already far the more important of the two, economically. Without it we should starve pretty quickly because without it the workmanship of certainty would cease, and only by way of that is mass-production possible. The productive branch on the other hand is declining, and in the course of the next two or four generations it may well have become economically negligible as a source of useful products. But, though, after that, the workmanship of risk may never again provide our bread, it may yet provide our salt. It will no doubt provide our space-craft too, and our more enormous scientific instruments.

The term 'crafts', that sadly tarnished name, may perhaps be applied to the part of the productive workmanship of risk whose justification is aesthetic, not economic (and not space-exploratory or particle-pursuing). The crafts on that definition will still have a slight indirect economic importance, in that they will enable designers to make relatively expensive experiments which the workmanship of certainty will deny them, and also to try out materials it denies them. But economics alone will never justify their continuation.

The crafts ought to provide the salt—and the pepper—to make the visible environment more palatable when nearly all of it will have been made by the workmanship of certainty. Let us have nothing to do with the idea that the crafts, regardless of what they make, are in some way superior to the workmanship of certainty, or a means of protest against it. That is a paranoia. The crafts ought to be a complement to industry (see preface).

For the crafts, in the modern world, there can be no half measures. There can be no reason for them to continue unless they produce only the best possible workmanship, free or regulated, allied to the best possible design: in other words, unless they produce only the very best quality. That quality is never got so quickly as more ordinary qualities are. The best possible design is seldom the one which is quickest to make, or anything like it; and, even where it is, the best quality of workmanship can usually be achieved only by the workman spending an apparently inordinate amount of time on the job. There are exceptions. Pottery, some hand-loom weaving and some jewellery, for instance, can be produced relatively cheaply. Moreover, in pottery at least, industry offers no serious competition, since the aesthetic qualities of 'studio pottery' are as yet rarely attempted in industrial production. Consequently these crafts flourish—though too seldom they produce the very best quality, or the best design—and people are making a reasonable living at them. But they are exceptions. The rule is, and always was, that the very best quality is extremely expensive by comparison with things of ordinary quality.

It is very probable that most people are beginning now to associate the word 'crafts' simply with hairy cloth and gritty pots. It is not quite realized perhaps that modern equivalents of the multitude of other kinds of workmanship we see in museums could and should be made: nor how astronomically expensive many of them would be.

Now the crafts, even when they do produce the very best quality, are in direct competition with producers of ordinary quality. The crafts are in no way comparable to the fine arts, a separate domain: far from it! The crafts are a border-ground of manufacturing industry, and early every object they make has its counterpart and competitor in something manufactured for the same purpose. In all but a very few trades exceedingly high quality is the last remaining ground on which the crafts can now compete.

Two of the fundamental considerations which will shape the future of the crafts are the time they must take over their work and the competition they must face. The differential in price between a product of craft, of the best quality, and a product of manufacture varies, naturally, according to the trade; but it is always large and sometimes huge. It ought to be and must be. Unless it is, the craftsman has no hope of anything approaching a modest professional standard of living, and he will never be able to command a better living than that.

The crafts will therefore survive as a means of livelihood only where there is a sufficient demand for the *very best quality at any price*.

That sort of demand still exists in some trades. *Haute couture* flourishes. Certain musical instruments, yachts, guns, jewellery, tailoring, and things of silver, are still in that kind of demand. But the demand is not large, by comparison, for instance, with the demand for contemporary paintings, or for antiques, at comparable prices. The situation of the craftsmen who make these things of the best quality is evidently precarious. The West End tailors and bootmakers are not finding it easy to exist any more.

In other fields that kind of demand has very nearly ceased in Britain. Cabinet- and chair-making, blacksmith's work, carving, hand-tool making, are examples. These are all cases where the differential is very large. Here the potential buyers have turned to antiques or else spend their money on things of other kinds.

It is not always clear why the demand has persisted in some fields but not in others. We may suspect that where it does persist the reasons are not always very creditable ones. But we need not concern ourselves with that, for it is absolutely certain that no demand for the best quality at any price can be re-created, or stimulated where it still persists, until it becomes a fact that a fair amount of work of that quality is being done and can be had.

Now, considering the time that is needed to do it, how can such work be made? It is obvious that it must be done, at first and for a long while afterwards, for love and not for money. It will have to be done by people who are earning their living in some other way.

It is sometimes hoped that a man can set up as, say, a cabinet-maker and aim at making a few pieces of the very best quality each year, so long as he keeps himself solvent by making other furniture to order, or for sale in competition with the manufacturers. This can be done and is being done. Some good furniture is being made in this way, but very, very little of the very best. The man who does it is likely to find that to make a moderate living he has to become a manager more than a maker—sales manager, works manager, despatch manager, buyer and accountant, as well as secretary, all rolled into one. Whatever he does of the very best quality will have to be done as a side line, very likely at week-ends. It will not increase proportionately to the other. If it were not for being his own master he might about as well make his living working in some other office or at some other trade, and make his two or three pieces of the very best quality in his spare time.

That is the logical conclusion. With certain exceptions, some of them precarious, the crafts, like the fine arts, are not fully viable. Only a very small proportion of painters can make enough money, by painting alone, to bring up a family, and that in a time when there is a climate of educated opinion very favourable to painting, a great international trade in contemporary paintings and a whole apparatus of distribution specifically for them: and when, above all, high prices for them are paid. None of these advantages is yet available to the crafts. Moreover, they are under a disadvantage which the painters are free from: the pressure of competition just mentioned.

Nearly all craftsmen, as nearly all painters and poets already do, will have to work part-time, certainly in the opening years of their career. One of the best professional cabinet-makers in Britain, Ernest Joyce, started as an amateur and learnt his job at first from books. 'Amateur', after all, means by derivation a man who does a job for the love of it rather than for money, and that happens also to be the definition, or at least the prerequisite, of a good workman. There is only one respect in which a part-time professional need differ from a man who can spend his whole working life at the job. He who works at it part-time must be content to work more slowly in his early years. Constant practice gives a certainty quite early in life which takes much longer to attain if one is working intermittently. Until he does attain it he must make up for the want of it by taking extra care and therefore extra time. In consequence his output will necessarily be very small; but that is unimportant. The only reason for doing this work is quality not quantity.

No one will find the patience to become a proficient workman of this sort unless he has a lively and continual longing to do it, and, given that, ways of learning the job will be found. There are books, there are examples of the work, and there are workmen. With the help of all these and with practice he will learn to do work of the highest standard. I doubt whether there is anything which a determined part-time professional could not attain to, except speed, and even that comes in time.

It is still commonly believed that a man cannot really learn a job thoroughly unless he depends on it for his living from the first and gets long experience at it. It is untrue. Two minutes experience teach an eager man more than two weeks teach an indifferent one. A man's earning hours and his creative hours can be kept separate and it may be that they are better separated. Painters and poets separate them. Are painting and poetry really so much easier than craftsmanship? Part-time seamen are making ocean voyages in small craft which any professional seaman of the days of sail would have highly respected. Is not that a parallel case? Astronomy, to take but one other example, has owed an immense debt to amateur observers and telescope makers from Newton and Sir William Herschel onwards. No one in that science would subscribe much to the idea that amateurs are apt to be amateurish. It is high time we separated the idea of the true amateur—that is to say the part-time professional—from the idea of 'do-it-yourself' (at its worse end) and all that is amateurish. The continuance of our culture is going to depend more and more on the true amateur, for he alone will be proof against amateurishness. What matters in workmanship is not long experience, but to have one's heart in the job and to insist on the extreme of professionalism.

That this kind of workmanship will be in the hands of true amateurs will be a healthy and promising state of affairs, not a faute de mieux, for if any artist is to do his best it is essential that his work shall not be influenced in the smallest degree by considerations of what is likely to sell profitably. What concerns us is the very best. It is that which must somehow be continued because the aesthetic quality of it is unique, and the tradition of it must be kept alive against a time when it will put out some new growth. The part-time professional will be in a position to do the very best even though he can turn out very little of it, and even though at first he will have to sell it at a price which pays him very little for his time. Why not? Whom will he be undercutting? Will there be placards saying 'Craftsmen Unfair to Automation'? That can't be helped.

Along this road there will still be pitfalls. The crafts and craftsmen have been bedevilled, ever since Ruskin wrote, by a propensity for striking attitudes. The attitude of protest I have mentioned already. Another one is the attitude of sturdy independence and solemn purpose (no truck with part-time workers: they are all amateurs; social value; produce things of real use to the community); another is the attitude of holier-than-thou (no truck with machinery; no truck with industry; horny-handed sons of toil; simple life, etc.). Another is the snob attitude, learnt from the 'fine' artists (we who practise the fine crafts are not as other craftsmen are). These are ridiculous nonsense by now, but who has not felt sympathy with them, all but the last, at one time or another? For nostalgia is always in wait for us. The workmanship of risk *was* in many ways better in the old days than it is now, there is no sense in pretending otherwise. Moreover, many of the trades we ought to set ourselves to continue are already taking the complexion of survivals from an older world. That should not prevent us from looking ahead. We must think of the future more than the past. Some trades which are dead economically are all alive in human terms, and still have much to show the world.

It remains to notice the most disastrous illusion which was encouraged by Ruskin's chapter, whether he meant it to be or not; and which has done the most harm: the illusion that every craftsman is a born designer. There are no born designers. People are born with or without the makings of a designer in them, but the use of those talents is only to be learnt very slowly by much practice. Any untrained but gifted man can knock up something which looks more or less passable as a design but the best design for industry is done by people who have really learnt their job; and it looks like it. The crafts are always liable to comparison with industry and they cannot afford to come off second best in design as well as in price.

Design is so difficult to learn now simply because the arts are in a state of violent flux and because there are great interests vested in constant innovation. There is no settled tradition. If there were, the profession would be far more quickly learnt. If the crafts develop as I envisage, perhaps few craftsmen will be able to go through a designer's training, but surely there will be designers who will work for them, and be glad of the chance even if they make no money by it at all. There will have to be an alliance between the craftsmen and the designers.

Some things, of course, can only be designed, or at any rate designed in detail, by the workman himself. Writing and carving are obvious examples. Other things, such as musical instruments, ought to go on being made to traditional designs (not 'reproduction' designs, which are quite a different thing. Tourte's pattern of violin bows has been in use ever since he evolved it: it is not a mere revival of something which had died out).

The whole future of the crafts turns on the question of design. If designers will only come to recognize it, the crafts can restore to them what the workmanship of certainty in quantity-production denies them: the chance to work without being tied hand and foot by a selling price: the chance to design in freedom. There is nothing more difficult or more necessary for the modern designer to attempt.

If the crafts survive, their work will be done for love more than for money, by men with more leisure to cultivate the arts than we have. Some of them will become designers, some not: that is not important: a designer is one sort of artist, a workman another. Instrumentalists do not feel any sense of inferiority because they are not composers. But the scale of what craftsmen could achieve by concerting their efforts, and the opportunity it would give designers, would be something not dreamt of. Cathedrals were built, if not with joy in the labour (*pace* Morris), quite certainly by concerted effort unaided by any plant to speak of but what the workmen made themselves. People are beginning to believe you cannot make even toothpicks without ten thousand pounds of capital. We forget the prodigies one man and a kit of tools can do if he likes the work enough. And, as for those trades by the workmanship of risk which do need plant, it is not impossible to imagine that associations of workmen will set up workshops by subscription.

The great danger is that spurious craftsmen, realizing that the workmanship of certainty can beat anyone at high regulation, will take to a sort of travesty of rough workmanship: rough for the sake of roughness instead of rough for the sake of speed, which is rough workmanship in reality. This can be seen already in some contemporary pottery.

One rather feels that painting, whatever else it does nowadays, has to take care to look as different as possible from coloured photographs. Have the crafts got to take care to look as different as possible from the workmanship of certainty? If that is the best aim they can set themselves, let them perish, and the quicker the better! If they have any sense of their purpose they will look different, right enough, without having to stop and think about it. It is infinitely to be hoped that free and rough work will continue, but not in travesty. One works roughly in order to get a job done quickly, but all the time one is trying to regulate the work in every way that care and dexterity will allow consistent with speed.

Free workmanship is one of the main sources of diversity. To achieve diversity in all its possible manifestations is the chief reason for continuing the workmanship of risk as a productive undertaking: in other words for perpetuating craftsmanship. All other reasons are subsidiary to that one, for there is increasingly a vacuum which neither the fine arts nor industry and its designers are any longer capable of filling. The contemporary passion for anything old, for junk and antiques, is no doubt symptomatic. The crafts in their future role may yet fill the vacuum but only if craftsmen achieve some consciousness of what they are for, only if they will set themselves the very highest standards in workmanship, and only then if they attract the voluntary services of the best designers. Workmanship and design are extensions of each other.

12

Commentary on the plates

The photographs which follow this chapter have been chosen to demonstrate some of the points I have been trying to explain.

I hope that the plates may be looked at in sequence and the commentaries on them read consecutively. They fall roughly into three groups, each mainly concerned with one particular point, as follows:

Plates 1–10*b* have been chosen in order to demonstrate something of the difference between regulated and free workmanship, and to contrast the workmanship of risk with that of certainty. Plates 10*c*–22*a* will, it is hoped, make clear certain points about diversity. Plates 22*b*–31*b* are mainly concerned with the question of good and bad workmanship.

1 Drawing-room cabinet. Holland and Sons, 1868.
Crown copyright. Victoria and Albert Museum.

2 Top of beer can.

Both these objects are of regulated workmanship. In neither is there any noticeable approximation. The can is entirely a product of the workmanship of certainty but it is less highly regulated than the cabinet, which is a product of the workmanship of risk. The central rivet of the can, for instance, is unevenly buckled, and there is unevenness in the impression of the lettering. These elements of free workmanship enhance the appearance by contrasting with the more completely regulated elements elsewhere.

The metal is not highly polished and for this reason the raw sheared edge on the right-hand end of the tab does not produce any marked sense of equivocality.

The can is an excellent piece of workmanship. Anyone accustomed to doing regulated work by the workmanship of risk must feel something of a pang at throwing such a thing away, for to make it by the workmanship of risk would be an intensely difficult and very long job.

The cabinet was made at a time when the art of workmanship in cabinet-making, and no doubt other trades, stood at its zenith. The quality of it in the best Victorian furniture will never be surpassed. The eighteenth and twentieth centuries rarely equalled and seldom approached it. This was the kind of workmanship Ruskin and Morris were inveighing against (chapters 10 and 11). This is one of the sorts of quality that the crafts must continue. There is real danger that it will otherwise die out entirely.

3 'Argus 400' computer.
By courtesy of Ferranti Ltd.
4 Carving of olive branch.

The carving is of free workmanship entirely, but the computer also exhibits some of it, in the wiring at the near ends of the centre and left-hand units, and also nearer the middle of the right-hand unit. The rest of what can be seen is of highly regulated workmanship, and the six hundred micro-miniature circuits (not visible here) are about as extreme examples of it as could be found. The two contrasted kinds of workmanship do not strike us as incongruous, probably because the same vein of neatness, order and compactness is evident in both.

The carving has characteristics which are typical of many other examples. Often the movements of the tool can be traced individually. The background, which 'is meant to be' flat, is not. The tool has overrun when the carver trimmed up the profile of the second leaf from the bottom on the left-hand side and has left a mark on the background which he could easily have removed but did not wish to. It has all been left as it is because higher regulation would, unless at the hands of an extraordinary artist, make the rendering less lively.

It is a commonplace observation that a finished picture is apt to lack the freshness of the sketch on which it was based. Free workmanship is essentially of the nature of a sketch.

5 Quay, with bollards. Barra.
6 Rough-hewn billet, with chopping-block and side-axe.

In the bollards, the chopping-block and the billet we see rough workmanship. The quay on which the large bollard stands has been partly overthrown by the sea but it must have been rough enough before. It has grandeur and must have had it always. Little or none of that quality is here attributable to design; it comes from the workmanship and its setting. No working drawing for such a quay could have suggested any of it.

A broad-axe and its products are not often seen now but were very common when rough workmanship was the only means of cheap manufacture. The chopping block against which the billet is leaning is simply a barked and trimmed log. There is a special attraction about the contrast between the quality of the chopped facets and the natural surface of the wood, and this has been felt, I think, all over the world if one may judge from ethnographical collections. They are great repositories of free and rough workmanship in some of its most attractive forms.

7 Lid of an earthenware crock.
8 Ceramic insulator for an electric power line.

The workmanship of risk does not necessarily produce free work such as we see in the crock lid, nor does the workmanship of certainty invariably produce high regulation, though it usually does and has done so in the case of this insulator.

Their similarity in material, colour and glaze makes a comparison between the two objects particularly interesting, because aesthetically they are quite different in quality, and each in its own way is good. The insulator is diversified less boldly than the crock. On the small scale its only diversification comes from the slightly uneven flow of the glaze which causes a small ripple in the clean line of every reflection. The shape in itself contributes little, but it casts freely curving shadows and shows complicated reflections. These adventitious formal elements extend the diversity on to a larger scale also, corresponding to a considerably longer range.

In the crock lid the larger-scale diversity is, by the free workmanship, built in and not adventitious at all. The striations, the flecked surface, the potter's hand-print and the asymmetry of the handle each build it up a stage farther and leave us with a vivid impression of life and decision.

The crock lid will serve to emphasize that in very few techniques is the workmanship of risk found in a pure state. The potter's wheel is an exact shape-determining system: a mechanical affair. However badly a pot is thrown it will pass through a stage of being very nearly circular in plan. But throwing pots on a wheel is the workmanship of risk none the less for that, as anybody can convincingly demonstrate.

9*a* Cringle of a sail.
9*b* Hop-garden early in the season.

Trades using similar techniques but making different things often work to different standards of regulation. It is interesting to compare the sail-maker's work shown here with the sewing of a suit of clothes. The two holes to the right of the cringle are not essentially different from button-holes. They are neat and workmanlike according to the accepted standards of heavy sailmaking, and are, moreover, a good deal neater than they need be for the sake of strength and durability. But they are not what one expects to get from a West End tailor. The standard of workmanship in weaving and tailoring is reflected in the price of it. This sail had to be cheaper and therefore rougher; but rougher does not imply worse. Sailmaking has a beauty of its own.

A hop-garden is one of the many admirable pieces of workmanship which agriculture and forestry produce and which contribute much of its quality to the countryside we know. It is predominantly free workmanship but there has always been traditionally an insistence on regulation as well, in ploughing a field, for example, or in aligning these hop-poles.

10*a* A pair of small blacksmith's tongs. Enlarged.
10*b* A putative working drawing for the tongs.
10*c* Silver tobacco-box. Early twentieth century. Enlarged.

The tongs were forged in the usual way, by eye, without any drawing and indeed without taking any measurements. Except for shaping the rivet the only tools used were an anvil, hammer, punch, and hot-chisel: almost a case of the workmanship of risk in a pure form, though an anvil and hammer constitute a shape-determining system to some extent. The workmanship is pretty rough, while the drawing shows the ideal form to which it approximates.

It is the ideal form, and not necessarily the design, which the drawing shows. No smith in his right mind would have a design like that drawing in his mind's eye while forging a pair of tongs. If required, of course, he could make something very close to it, but any designer who required him to would be wasting money.

The photograph of the tobacco-box shows, though incompletely, the screen of minute scratches referred to in chapter 9.

The work of the engraver, such as shown here, is much a matter of interpreting. An engraved line never has the same quality as a drawn one, for it exists in three dimensions not two. Here, for example, there is a highlight next to each mark of the graver. The edge which was sharp at first has worn very slightly round and produced this effect.

After considerable use the box is beautifully diversified in a way typical of much silver, through the shading and reflections created by its pebble-like shape, through the quality of the engraved line, and through the quality of the soft faint mesh of dents and scratches in the metal.

11 Diffusion apparatus.
By courtesy of Ferranti Ltd.

12 Blowing engine, 1825–50 (model).
Crown copyright. Science Museum, London.

In plate 11 we see the quality of no-diversity at an extreme. It is the quality which has come to be called 'clinical' and which we associate with hospital and laboratory apparatus. Since it shows up dust and usually belongs to surfaces which are easy to wash clean, it is an excellent and desirable quality in apparatus of this sort. Whether it is appropriate to large agglomerations of large buildings is more than doubtful. It will come to seem infuriatingly vacuous.

The representation of the building which surrounds the model engine does look vacuous; and in its special context quite properly so: for it is no doubt intended as an unobtrusive foil to the model itself. The model engine looks 'real' but the model building does not, yet the model building is perfectly realistic in treatment. Their incongruity comes about simply because on the one hand the engine is highly diversified at every scale, while on the other the building is undiversified at any scale or very nearly so. Since in addition the engine looks all black and the building all white, the building all matt and the engine all shiny, the two become completely dissociated.

The beam of the engine and its ancillary parts are extraordinarily diversified and their appearance is delightful.

13 Maudslay's table engine (model).
Crown copyright. Science Museum, London.

14 Part of engraving of the same engine by G. Gladwin after J. Clement.

The engine and the engraving of it strike the eye very differently, for whereas the engine shows an extended range of diversity embracing many different qualities—compare the rim of the flywheel, the crankshaft end and its bearing, and the surface of the cast-iron drum below it—the engraving on the other hand shows a fairly restricted range and little variety: for all its different surfaces are represented alike by parallel straight lines differing only in width and interval. It is a very highly regulated piece of work.

We may confidently say the engraving, like the drawing in plate 10*b*, represents the ideal form but not the designer's intention; for Maudslay was an artist and perhaps the most notable workman of his day in this field. It is not to be doubted that his engines were made exactly as he intended they should be.

There is more, far more, in the appearance of the engine than in the engraving, and all of that has been added by workmanship, not by design. The engraving, like any working drawing, is merely a summary of all the directions a designer would be able to give about the engine's size and shape, both measurable. Beyond that he could give directions about the measurable properties of the materials to be used in making it, and the measurable strength of it when made. That is about as far as directions will go. When a designer tries to specify the quality of workmanship, in fact to say how it shall strike the eye, what he does, in effect, is to point to *samples of already existing workmanship* and say: 'Do it like that'.

15 The prophet Haggai. Giovanni Pisano. From Siena cathedral. Last quarter of thirteenth century.
Crown copyright. Victoria and Albert Museum, London.

16 Carver rack-cramp.

The head shows rough workmanship at its highest pitch of refinement. It is a masterly demonstration of the principle of diversity: that small elements barely at the threshold of recognition are capable of intensifying the character of the larger forms which underlie them. The head was evidently designed to be seen at some considerable distance, at which, if it had been more highly finished, its impact would have been less. No doubt its carver would have thought it quite unfit to be seen close by.

The design of the small elements has not been left to chance. The vermiform trenches were cut with a machine-tool but, although the running drill was, at its business end, in effect the same thing as a modern milling-cutter or router-bit, its drive must have been comparatively slow and every mark it made must have been a matter of deliberate intention.

The cramp also gains in appearance from the roughness of its surface when the individual accidents of it are too far away to be distinguishable. Something of this may be seen by putting the photograph at a distance. Although its diversity is mainly achieved by the workmanship of certainty the quality of the rivets and of the ground parts of the casting is partly attributable to the workmanship of risk. They have been finished well, and would have spoilt the job if they had not.

17 Screen by Sakai Hoitsu.
By permission of the trustees of the British Museum.
18 Ruined water-mill in a wood.

The stone walls of the mill have intruded on the trees and the stream. The silver leaf of which the screen's background is made has, by tarnishing, obtruded its pattern on the picture. The result in each case has been fortunate, because the masonry and the squares of tarnished silver are each widely and delicately diversified in a fashion which enhances the mood evoked by the surroundings: though how it does so is impossible to say. It is doubtful whether any agency but time could produce quite such happy effects. It is to effects like these that the Japanese term *sabi* refers.

19*a* Tail of De Havilland Heron aircraft.
19*b* Part of cast-iron fireplace.

Both these examples are diversified at a distance by reason of the gradation of shading which their curved shapes produce, and at close range by the quality of the surface of their paint.

The lettering on the aircraft supplies formal elements of intermediate size between the over-all shape and the minute incidents of the surface.

20*a* Number 27, St James's Street, London. Peter and Alison Smithson, Architects.

20*b* Great Shoesmiths Farm, Wadhurst, Sussex.
By courtesy of Mr and Mrs Fogden.

The elevations of this modern building are of an austere design yet it has an attractive dignity and calm about it. The stonework of the pillars is of moderately free workmanship, and this, with the character of the stone surfaces and their fossils, has lent it a diversity which prevents it from being forbidding.

The farm-house is of a sort whose distinction is often said to be due to fine proportions. But imagine a replica of it with every dimension faithfully copied, using, say, smooth opaque vitreous panels for the walls, plated steel or polished alloy sections for the windows, unpainted cement rendering for the chimneys, and regular, shiny, hard plain tiles for the roof: or imagine any other combination of components equally deficient in diversity and yet preserving all the proportions; and it will become evident that proportions are not enough. Workmanship matters quite as much as they do.

The materials of which a building is made, on the other hand, seldom have much importance independently of workmanship. For example, the same kind of clay, the same kind of timber, the same kind of stone as we see here, could each by wrong handling be made quite repellent. It is not the material but the kind of work put into it which counts. Each and every material there is can be made to look nasty easily enough.

21*a* Ramsden solar and scroll microscope. Late eighteenth century.
Crown copyright. Science Museum, London.

21*b* 'The blind Earl pattern'. Worcester porcelain, early nineteenth century. Enlarged.

The microscope is highly regulated and yet shows an extended range of diversity. This is attributable, as usual, partly to workmanship and partly to design. The lively ungeometric profile of the curving support contributes much, and the numerous mouldings turned in the brass, combined with the tool-finish visible throughout, diversify the surface by their light and shade and broken reflections. So do the larger cylindrical forms of the barrels. The engraving and the knurling on the screws are also important among the larger elements.

The painting on the porcelain dish is of astonishingly highly regulated workmanship, considering that it was done freehand, at speed, with a brush. It shows the pitch to which dexterity can be brought by continuous practice.

The photograph has been made to over-emphasize the painted lines, which are in gold, and the very delicate relief modelling within the leaves, most of which is lost to view at a range of about six feet and not all of which can ever be seen at once, so that it is an important diversifying element.

The workmanship has given this pattern an extraordinarily lovely quality.

22 *a* Bread-board showing marks of wear and scrubbing. Enlarged.
22 *b* Title page, 1675.

The delightful appearance which wear and scrubbing have given the elm breadboard is the outcome, not so much of one surface quality, as of a whole range of formal elements at different scales, running at one end into the domain of low-relief modelling and at the other into nuances far too subtle for this or any photograph to show. This of course is true of the majority of particular surface qualities, so called. They are the outcome of the diversity given by a wide variety of formal elements most of which are just at, or just below, the threshold of recognition at the distances at which we ordinarily handle things: so that the many diverse elements visually fuse together and give the surface its particular flavour.

In techniques such as printing and casting we sometimes see free or rough workmanship reproduced by the workmanship of certainty. This title page is an example. The printer has stood one of the '*N*'s on its head and the letters are badly out of line. The result has a certain charm but we feel that either he was not clever or that something lively must have been happening the evening before. We cannot believe it was meant to be like that.

The intentions, in printing, are clear to everyone. The letters are meant to be aligned. In the present case the situation is saved, just, by the fact that the letters are themselves of quite free workmanship. Each of the '*I*'s for example is quite a different shape, each is unsymmetrical and one is far out of square. If these had been modern highly regulated and exactly repeated letters their want of alignment would be bad workmanship unmistakably.

23*a* Front of a drawer. Early eighteenth century.
23*b* Back of the same drawer.

It must be said at once that much furniture of this date was better made than this contemptible example, but really bad workmanship was not uncommon and it may be that the best was rarer than is generally supposed. Much of what survives is quite well made, but that perhaps merely confirms that quality is a good preservative. Things that are well made and well designed tend to survive and things like this drawer tend to perish; which indeed it would probably have done but for the walnut veneering on the front, of which the workmanship is quite fair, and to which age has given a pleasant quality.

At the back of it the point of the hand-forged nail has been shamelessly airing itself for two hundred and fifty years or so, but the slip of wood below the bottom at the corner, and the slobber of glue above it, are more recent embellishments.

It is of very bad workmanship indeed, but that has not prevented it from serving its purpose all those years. The nail with its point in the air has undeniably kept the bottom on the drawer. Workmanship is not to be judged on a merely utilitarian footing.

24*a* Dovetailing: on the left, of a late Victorian drawer; on the right, of the same drawer as in plates 23*a* and *b*.
24*b* Well-fitted dovetails of about 1960. Enlarged.

The eighteenth-century drawer has been repaired—fittingly—with a couple of wire nails. Its dovetail joint had failed, being extremely badly cut. The excellent wainscot oak of which the drawer side is made deserved a better fate.

The Victorian drawer shows a respectable standard of workmanship in a job of moderate price. These joints were sawn by hand, usually at great speed: very much the workmanship of risk. It is a point of style in hand dovetailing drawers in England that the wedge-shaped 'pins' shall be slim and quite symmetrical. The joint would be just as strong if they were neither.

The dovetails in plate 24*b* are of more highly regulated workmanship, and they fit exactly; but they were probably a very slow job compared with the Victorian example.

25 Memorial, Beddingham, Sussex.
26 New doorway.

The memorial itself, and particularly the lettering on it, are of quite highly regulated workmanship. The flint wall in which it is set is rough. There is no sense of incongruity because the memorial is clearly a self-contained object, distinct from the setting. It would never occur to us for a moment that the flint wall was rough 'by mistake' because of bad workmanship.

Now, in the doorway we have, it would seem, much the same state of affairs: rough workmanship in the wall, regulated work in the door for which it makes a setting. Yet in this case there is evident bad workmanship.

The two cases are not in fact comparable. The workmanship of the doorway is bad for these reasons: The wall is of rough workmanship but not consistently, for the lintel to the right is quite highly regulated. Yet to the left the lintel becomes partly rough. What was the intention? Rough work or regulated work? We are left in doubt. But, since the components of the door-frame and also the door itself are regulated, and since the lintel, on the right at least, corresponds to them, we feel that the roughness of the left-hand end of the lintel cannot have been intended but must be the result of bad workmanship. And, considering the door and frame, the component parts of the frame are evidently quite straight, quite flat and quite parallel-edged, with no sign of approximation about them. But the way they fit together is very approximate and the inequality in the width of the gap beside and above the frame is approximate too. We feel that decidedly it was not meant to be like that and we recognize it as bad workmanship.

But the doorway will hold together for many years to come in spite of that.

27 Stile.
28 Kilmory Cross, Argyllshire.

In a thing as weathered and time-worn as this stile we think open joints and partial collapse are honourable scars, but if these same timbers when freshly sawn had been left to stand in the same way, with the joints nearly falling apart, we should consider it gross bad workmanship. Our judgement is not simply a matter of sentiment. Age and weather have converted the originally regulated workmanship of the rails, post and spur into an evidently rough approximation: a very rough one. So we find the open joints perfectly in key with the rest of the workmanship—the workmanship of wind and wet. Because it is consistent we half believe it intentional, or at least regard it as if it had been.

There can be no question that the cross was made asymmetrical deliberately, and not because of bad workmanship. The complete regulation of the shaft and the roundel is proof of the carver's competence, while the careful finish shows that the odd profile of the right arm of the cross cannot have been overlooked in haste. The asymmetry must therefore have had a symbolic value; and it is so obvious and unabashed that it convinces us that it was indeed intentional. Its abruptness is a little softened by the asymmetry of the pattern carved within the roundel and the arms.

29 Wooden dish.
30 Mass-produced glass jar.

The large wooden dish shows free workmanship throughout. In one plane the tool has been partly guided by a jig, but not completely controlled. The intersecting flutes are thus uneven in width and depth and the pattern is all moderately irregular. One feels no disquiet about that. The irregularity is evidently intentional. But one of the flutes, which runs from the point where the dish touches the table upwards in a direction as it were a little to the east of north, has been cut a trifle too deep. In consequence it is too prominent. This, just as evidently, was not intended, and it slightly spoils the effect by interrupting the drift of the pattern.

The jar illustrates the fact that with the workmanship of certainty in mass-production an extended scale of diversity is likely to be achieved more easily where transparent or translucent materials are in use. The inner surface of the glass is unregulated and only the outer form of it has been determined by a mould. The wandering inner surface distorts and modifies the reflected lights and imparts diversity.

In the rather rare cases where diversity comes easily to it, the workmanship of cheap mass-production can be very beautiful; as we see here.

31*a* Viola bows.
By courtesy of the maker, Mr A. R. Bultitude.
31*b* Handles of tools. Enlarged.

The bow draws much of its quality from the carving of its tip and from the highly regulated juxtaposition, by inlaying, of different materials which fit exactly together and are rubbed down so as to present a single unbroken surface. Thus any specific quality in the material is made to speak for itself, unaided by relief or any other emphasis. The different materials take different degrees of polish from the same rubbing and so are differentiated by the play of light as well as by their inherent qualities and different colours.

Bows of this standard probably represent the most exquisitely regulated work being done in wood at the present time.

The intended design of all the three handles of tools is evidently what we see at the top example, which is admirably managed. The one in the middle has gone a little wrong at the left-hand end, while the one at the bottom is bad altogether—in appearance. As regards usefulness it is just as good as the others. One's fingers cannot tell the difference.

Plates

1 Drawing-room cabinet

2 Top of beer can

3 'Argus 400' computer

4 Carving of olive branch

5 Quay, with bollards. Barra

6 Rough-hewn billet

7 Earthenware crock-lid

8 Ceramic insulator

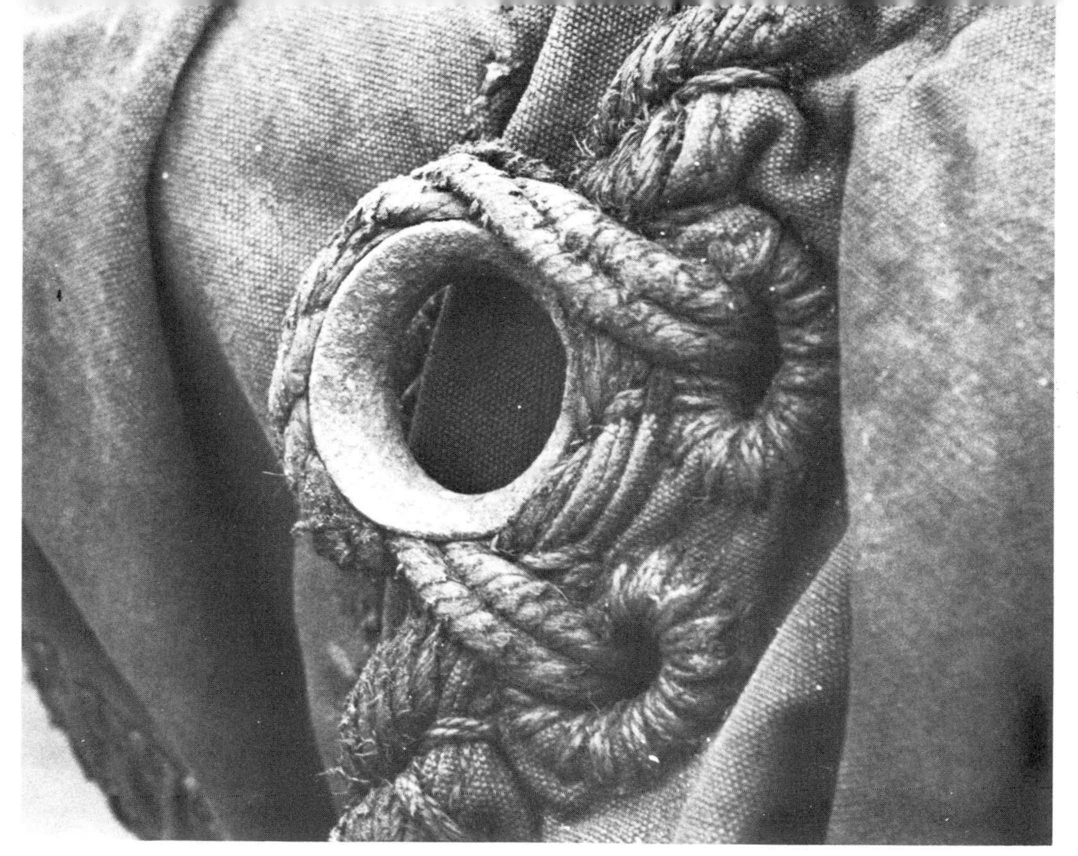

9*a* Cringle of a sail

9*b* Hop-garden

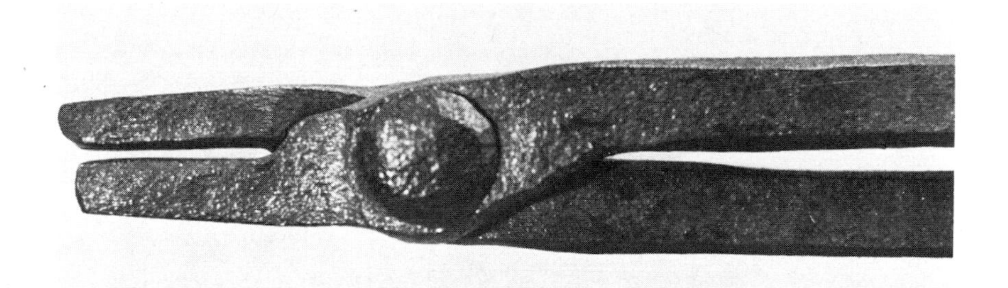

10*a* Blacksmith's tongs

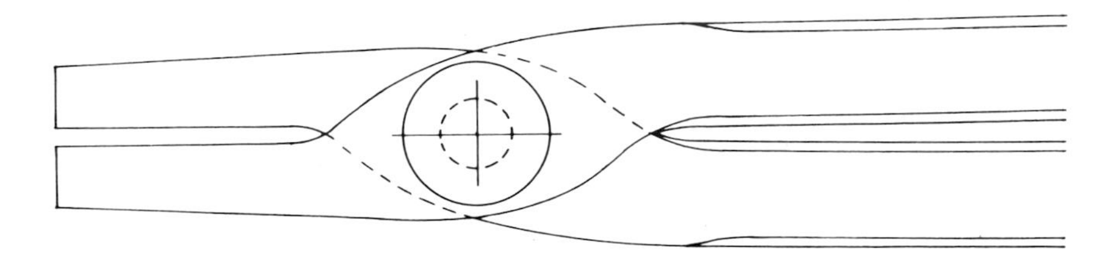

10*b* Working drawing of the tongs

10*c* Silver tobacco-box

11 Diffusion apparatus

12 Blowing engine

13 Maudslay's table engine

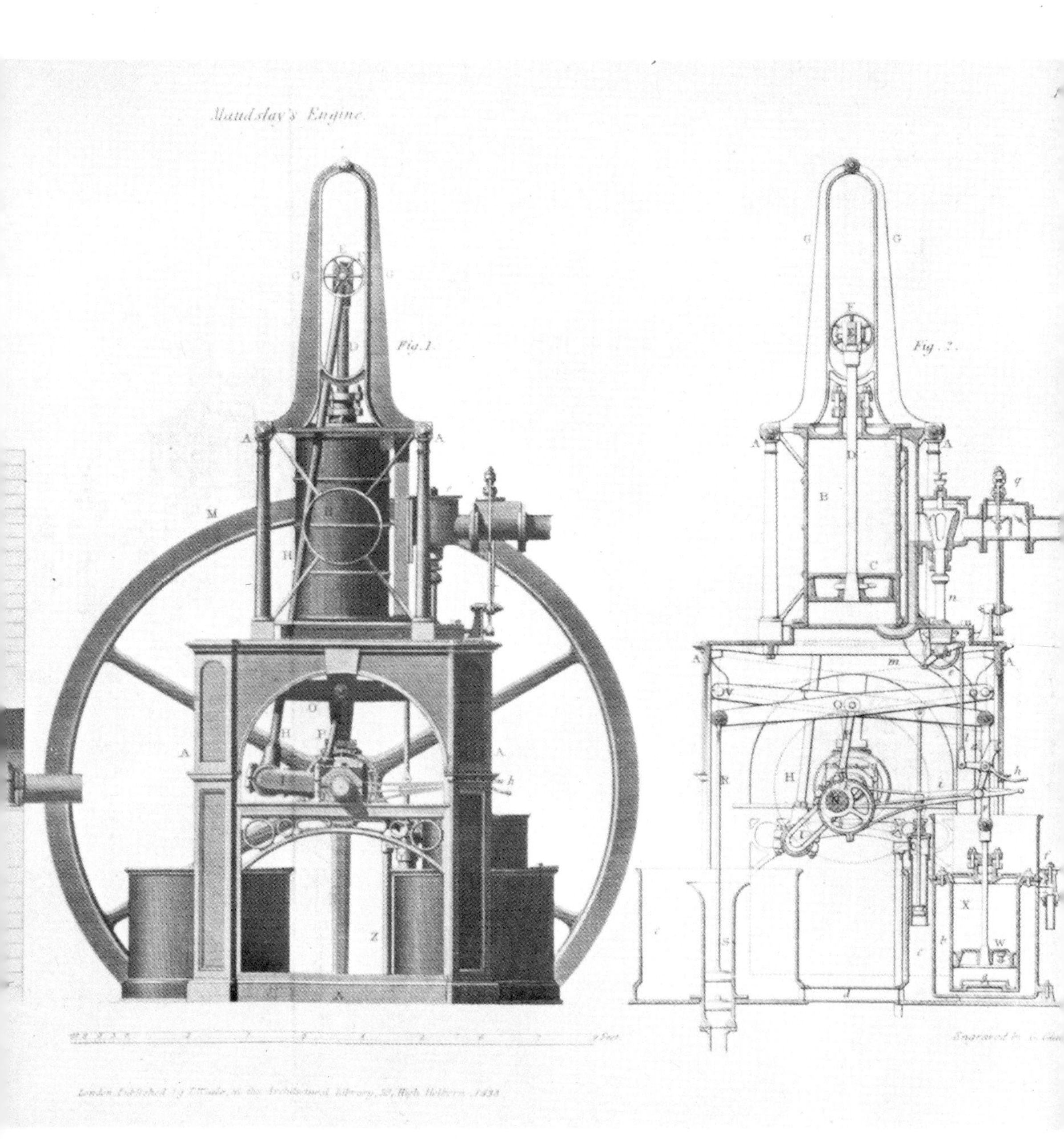

14 Engraving of the same engine

15 The prophet Haggai. Pisano

16 Carver rack-cramp

17 Screen by Sakai Hoitsu

18 Ruined water-mill

19*a* Tail of aircraft

19*b* Part of cast-iron fireplace

20*a* No. 27, St James's Street, London

20*b* Great Shoesmiths Farm, Wadhurst

21*a* Ramsden solar and scroll microscope

21*b* 'The blind Earl pattern'

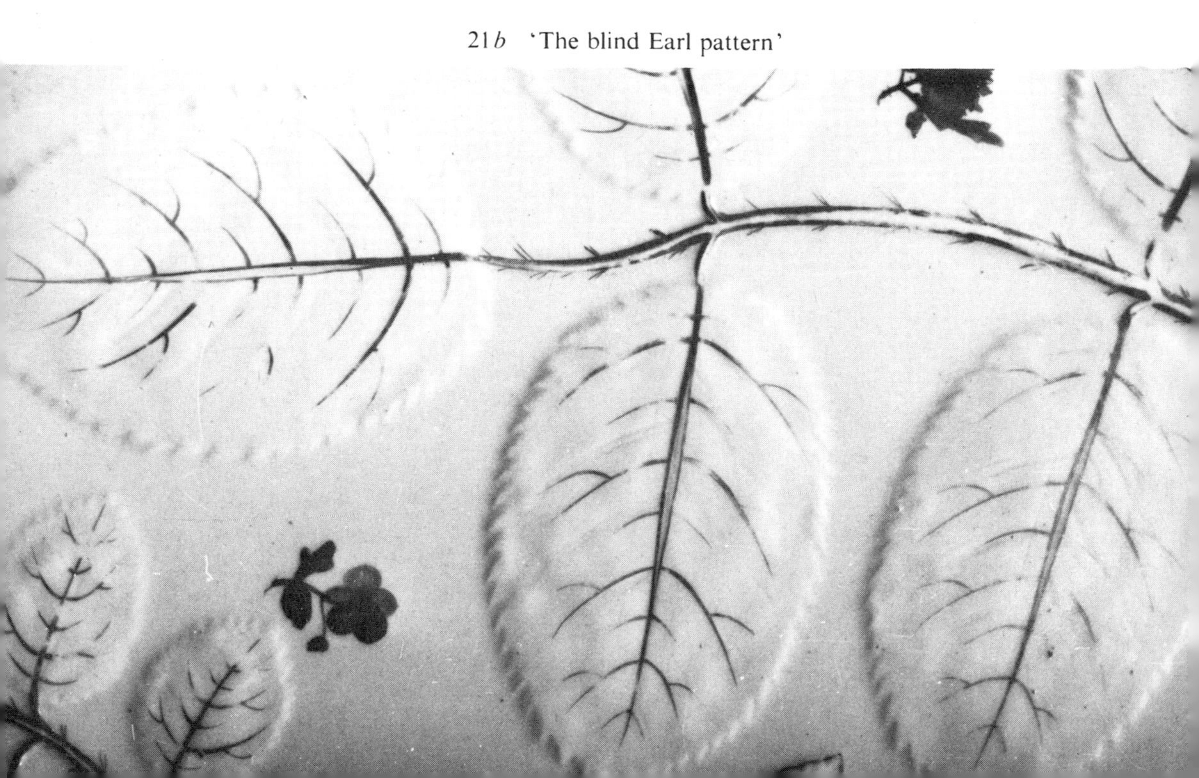

22*a* Bread-board

22*b* Title page

The Art of Painting

IN

MINITURE

OR

LIMNING:

23*a* Front of a drawer

23*b* Back of the same drawer

24*a* Dovetailing

24*b* Well-fitted dovetails

25 Memorial

26 New doorway

27 Stile

28 Kilmory Cross, Argyllshire

29 Wooden dish

30 Glass jar

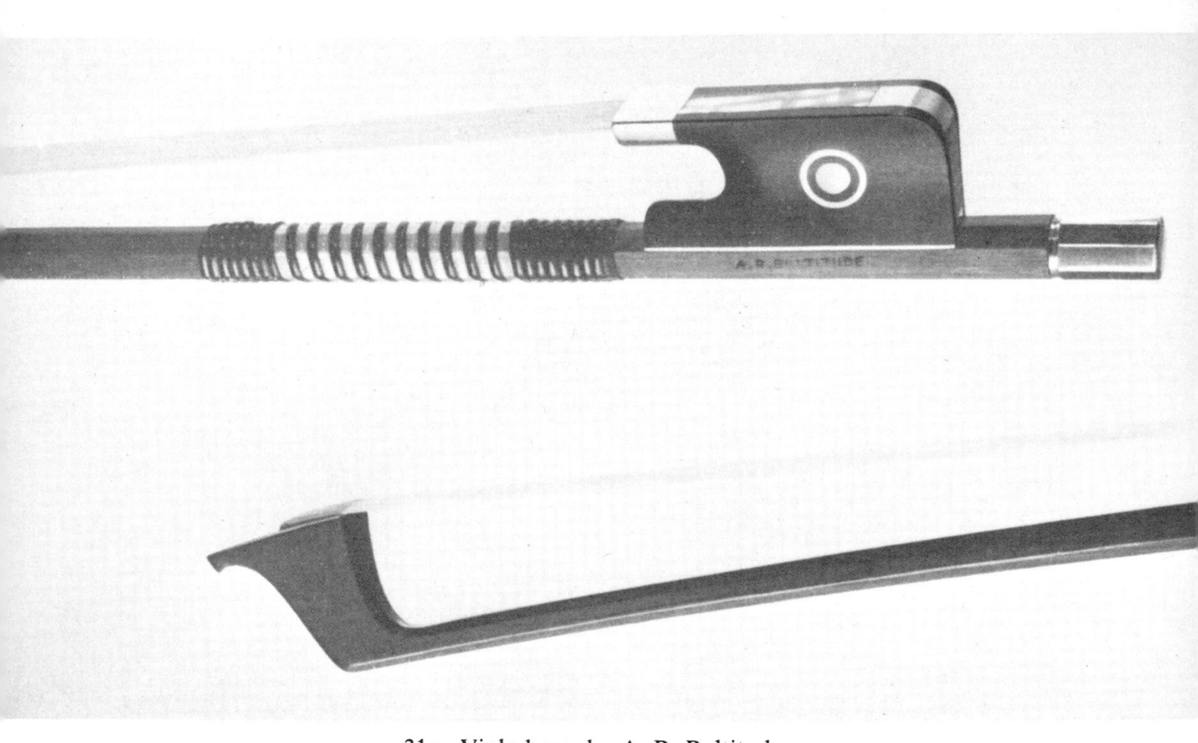

31*a* Viola bows by A. R. Bultitude

31*b* Handles of tools

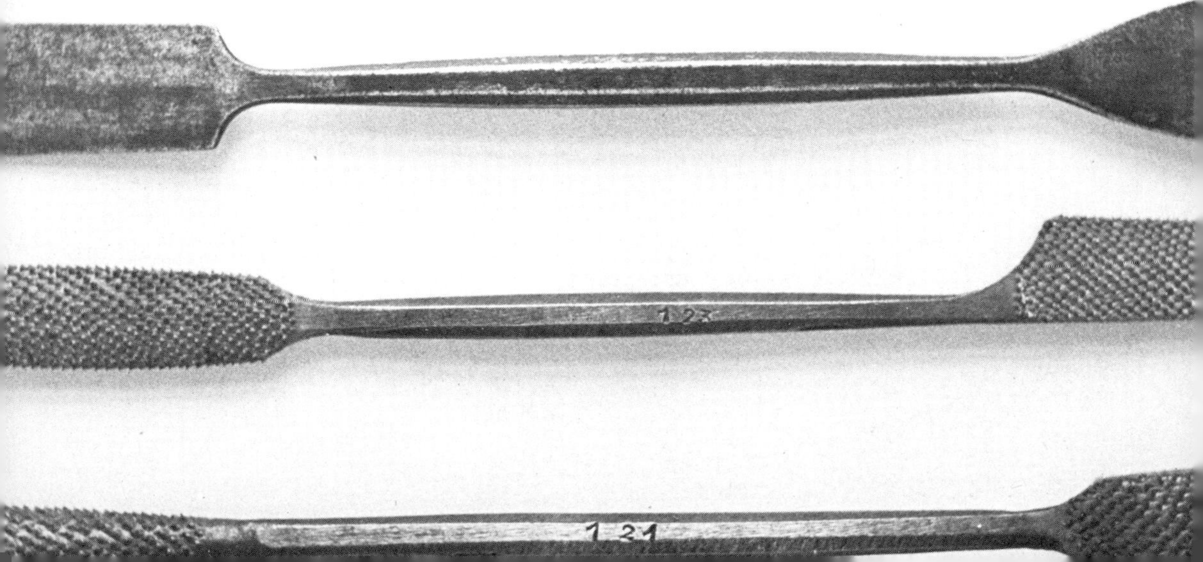

Index

Index